Random Walks in Biology

Random Walks in Biology
Expanded Edition

Howard C. Berg

Princeton University Press

Published by Princeton University Press, 41 William Street,
Princeton, New Jersey 08540
In the United Kingdom: Princeton University Press, Chichester,
West Sussex

Copyright © 1983 by Princeton University Press; expanded edition
copyright © 1993 by Princeton University Press

All Rights Reserved

Library of Congress Cataloging-in-Publication Data

Berg, Howard C., 1934–
   Random walks in biology / Howard C. Berg. — Rev. ed.
     p. cm.
   Includes bibliographical references (p.) and index.
   ISBN 0-691-00064-6 (pb)
   1. Biometry. 2. Random walks (Mathematics) I. Title.
QH323.5.B45   1993
574'.01'519282—dc20   93-12708

First Princeton Paperback printing, expanded edition, 1993

Princeton University Press books are printed on acid-free paper
and meet the guidelines for permanence and durability of the
Committee on Production Guidelines for Book Longevity of the
Council on Library Resources

Printed and bound by CPI Group (UK) Ltd, Croydon, CR0 4YY

ISBN-13: 978-0-691-00064-0 (paperback)

ISBN-10: 0-691-00064-6 (paperback)

Table of Contents

Preface to the Expanded Edition	ix
Preface (1983)	xi
Introduction	3

Chapter 1. **Diffusion: Microscopic Theory** — 5

One-dimensional random walk	6
Two- and three-dimensional random walks	11
The binomial distribution	12
The Gaussian distribution	14

Chapter 2. **Diffusion: Macroscopic Theory** — 17

Fick's equations	17
Time-dependent solutions to Fick's equations	21
Steady-state solutions to Fick's equations	25

Chapter 3. **Diffusion to Capture** — 37

Probability of capture	38
Mean time to capture	42

Chapter 4. **Diffusion with Drift** — 48

Random walk with drift	48
Fick's equations for diffusion with drift	50
Viscous drag	51
Sedimentation rate	58
Electrophoresis	62

Chapter 5. **Diffusion at Equilibrium** — 65

Derivation of the Boltzmann equation	68
The importance of kT	70
Mean-square velocity	71
Einstein-Smoluchowski relation	71

Equilibrium sedimentation ... 72
Density-gradient sedimentation ... 73
Isoelectric focusing ... 74

Chapter 6. Movement of Self-propelled Objects ... 75

Life at low Reynolds number ... 75
Flagellar propulsion ... 78
Motility of *Escherichia coli* ... 81
Rotational diffusion ... 81
Random changes in direction ... 85
Poisson statistics ... 86
Bacterial diffusion ... 93

Chapter 7. Other Random Walks ... 95

Countercurrent distribution ... 95
Partition chromatography ... 100
Sedimentation field-flow fractionation ... 102
Autocorrelation analysis ... 105

Appendix A. Probabilities and Probability Distributions ... 111

Probabilities ... 111
Probability distributions ... 113
The binomial distribution ... 118
The Gaussian distribution ... 121
The Poisson distribution ... 123

Appendix B. Differential Equations ... 125

Ordinary differential equations ... 125
Partial differential equations ... 129
Numerical solutions ... 131

Appendix C **Addendum to Chapter 6**	**134**
Strategies for chemotaxis	134
Movement of cells across artificial membranes	136
Movement of ions across cell membranes	139
Appendix D **Constants and Formulas**	**143**
Bibliography	145
Index	149

Preface to the Expanded Edition

I have added an appendix in which more is said about the behavior of self-propelled objects, the subject of Chapter 6. This includes the movement of ions across cell membranes. Ions are not self-propelled, but they drift in electric fields generated by their own displacement. Also, I have corrected the errata and updated the bibliography. Otherwise, this edition is identical to the first. I thank Joel E. Cohen, Nicholas J. Cox, Amal K. Das, Markus Meister, and Katsuhisa Tawada for spotting mistakes and offering suggestions. Finally, I wish to pay my respects to George Pólya, who coined the term random walks and initiated their study. He did this in 1921 in a paper in which he proved that a randomly moving point always returns to its initial position in one or two dimensions but not in three or more. He died 7 September 1985.

Howard C. Berg
Cambridge, Massachusetts
29 November 1992

Preface

This book grew out of lectures given in courses on biochemistry and biophysics at Harvard, at the University of Colorado in Boulder, and at Caltech. It is offered, in part, with the conviction that biologists and biochemists would enjoy their work more if they thought less in terms of thermodynamics and learned some of the rudiments of statistical physics. The intent is pedagogical, not seminal. Statistical phenomena of the sort treated here are met daily in the laboratory but are rarely appreciated as such.

Much of what I know about the physics of random walks has been learned through association with Edward M. Purcell. I have drawn on our conversations, correspondence, and joint publications. I am particularly grateful for his comments on the manuscript. I also wish to thank Steve Block, Rick Lapidus, Markus Meister, and Jeff Segall for help with the exposition and the physics and Connie Katz for masterful word processing. Work on the motile behavior of bacteria, described in Chapter 6, was supported by grants from the Research Corporation, the National Science Foundation, and the National Institutes of Health.

Without doubt, this book can be improved. If you have comments or suggestions, I would be pleased to have them.

Howard C. Berg
Pasadena, California
29 November 1982

Random Walks in Biology

Introduction

Biology is wet and dynamic. Molecules, subcellular organelles, and cells, immersed in an aqueous environment, are in continuous riotous motion. Alive or not, everything is subject to thermal fluctuations. What is this microscopic world like? How does one describe the motile behavior of such particles? How much do they move on the average? Questions of this kind can be answered only with an intuition about statistics that very few biologists have. This book is intended to sharpen that intuition. It is meant to illuminate both the dynamics of living systems and the methods used for their study. It is not a rigorous treatment intended for the expert but rather an introduction for students who have little experience with statistical concepts.

The emphasis is on physics, not mathematics, using the kinds of calculations that one can do on the back of an envelope. Whenever practical, results are derived from first principles. No reference is made to the equations of thermodynamics. The focus is on individual particles, not moles of particles. The units are centimeters (cm), grams (g), and seconds (sec).

Topics range from the one-dimensional random walk to the motile behavior of bacteria. There are discussions of Boltzmann's law, the importance of kT, diffusion to multiple receptors, sedimentation, electrophoresis, and chromatography. One appendix provides an introduction to the theory of probability. Another is a primer on differential equations. A third lists some constants and formulas worth committing to memory. Appendix A should be consulted while reading Chapter 1 and Appendix B while reading Chapter 2. A detailed understanding of differential equations or the methods used for their solution is not required for an appreciation of the main theme of this book.

Chapter 1

Diffusion: Microscopic Theory

Diffusion is the random migration of molecules or small particles arising from motion due to thermal energy. A particle at absolute temperature T has, on the average, a kinetic energy associated with movement along each axis of $kT/2$, where k is Boltzmann's constant. Einstein showed in 1905 that this is true regardless of the size of the particle, even for particles large enough to be seen under a microscope, i.e., particles that exhibit Brownian movement. A particle of mass m and velocity v_x on the x axis has a kinetic energy $mv_x^2/2$. This quantity fluctuates, but on the average $\langle mv_x^2/2 \rangle = kT/2$, where $\langle \ \rangle$ denotes an average over time or over an ensemble of similar particles. From this relationship we compute the mean-square velocity,

$$\langle v_x^2 \rangle = kT/m, \tag{1.1}$$

and the root-mean-square velocity,

$$\langle v_x^2 \rangle^{1/2} = (kT/m)^{1/2}. \tag{1.2}$$

We can use Eq.1.2 to estimate the instantaneous velocity of a small particle, for example, a molecule of the protein lysozyme. Lysozyme has a molecular weight 1.4×10^4 g. This is the mass of one mole, or 6.0×10^{23} molecules; the mass of one molecule is $m = 2.3 \times 10^{-20}$ g. The value of kT at 300°K (27°C) is 4.14×10^{-14} g cm^2/sec^2. Therefore, $\langle v_x^2 \rangle^{1/2} = 1.3 \times 10^3$ cm/sec. This is a sizeable speed. If there were no obstructions, the molecule would cross a typical classroom in about 1 second. Since the protein is not in a vacuum but is immersed in an aqueous medium, it does not go very far before it bumps into molecules of

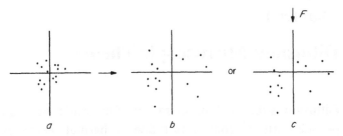

Fig. 1.1. Particles confined initially in a small region of space (*a*) diffuse symmetrically outward (*b*) or outward and downward (*c*) if subjected to an externally applied force, *F*.

water. As a result, it is forced to wander around: to execute a random walk. If a number of such particles were confined initially in a small region of space, as shown in Fig. 1.1a, they would wander about in all directions and spread out, as shown in Fig. 1.1b. This is simple diffusion. If a force were applied externally, such as that due to gravity, the particles would spread out and move downward, as shown in Fig. 1.1c. This is diffusion with drift. In this chapter, we analyze simple diffusion from a microscopic point of view. We look at the subject more broadly in Chapters 2 and 3. Diffusion with drift is considered in Chapter 4.

One-dimensional random walk

In order to characterize diffusive spreading, it is convenient to reduce the problem to its barest essentials, and to consider the motion of particles along one axis only, say the x axis, as shown in Fig. 1.2. The particles start at time $t = 0$ at position $x = 0$ and execute a random walk according to the following rules:

1) Each particle steps to the right or to the left once every τ seconds, moving at velocity $\pm v_x$ a distance

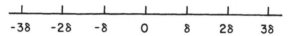

Fig. 1.2. Particles executing a one-dimensional random walk start at the origin, 0, and move in steps of length δ, occupying positions 0, $\pm\delta$, $\pm 2\delta$, $\pm 3\delta$, ...

$\delta = \pm v_x \tau$. For simplicity, we treat τ and δ as constants. In practice, they will depend on the size of the particle, the structure of the liquid, and the absolute temperature T.

2) The probability of going to the right at each step is 1/2, and the probability of going to the left at each step is 1/2. The particles, by interacting with the molecules of water, forget what they did on the previous leg of their journey. Successive steps are statistically independent. The walk is not biased.

3) Each particle moves independently of all the other particles. The particles do not interact with one another. In practice, this will be true provided that the suspension of particles is reasonably dilute.

These rules have two striking consequences. The first is that the particles go nowhere on the average. The second is that their root-mean-square displacement is proportional not to the time, but to the square-root of the time. It is possible to establish these propositions by using an iterative procedure. Consider an ensemble of N particles. Let $x_i(n)$ be the position of the ith particle after the nth step. According to rule 1, the position of a particle after the nth step differs from its position after the $(n-1)$th step by $\pm\delta$:

$$x_i(n) = x_i(n-1) \pm \delta. \qquad (1.3)$$

According to rules 2 and 3, the $+$ sign will apply to roughly half of the particles, the $-$ sign to the other half. The mean displacement of the particles after the nth step can be found by summing over the particle index i and

8—Diffusion: Microscopic Theory

dividing by N:

$$\langle x(n) \rangle = \frac{1}{N} \sum_{i=1}^{N} x_i(n). \qquad (1.4)$$

On expressing $x_i(n)$ in terms of $x_i(n-1)$, Eq. 1.3, we find

$$\langle x(n) \rangle = \frac{1}{N} \sum_{i=1}^{N} [x_i(n-1) \pm \delta]$$

$$= \frac{1}{N} \sum_{i=1}^{N} x_i(n-1) = \langle x(n-1) \rangle. \qquad (1.5)$$

The second term in the brackets averages to zero, because its sign is positive for roughly half of the particles, negative for the other half. Eq.1.5 tells us that the mean position of the particles does not change from step to step. Since the particles all start at the origin, where the mean position is zero, the mean position remains zero. This is the first proposition. The spreading of the particles is symmetrical about the origin, as shown in Fig.1.3.

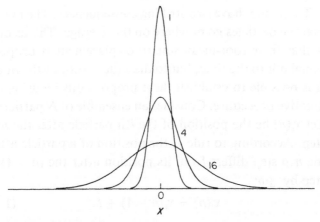

Fig. 1.3. The probability of finding particles at different points x at times $t = 1, 4,$ and 16. The particles start out at position $x = 0$ at time $t = 0$. The standard deviations (root-mean-square widths) of the distributions increase with the square-root of the time. Their peak heights decrease with the square-root of the time. See Eq. 1.22.

Diffusion: Microscopic Theory—9

How much do the particles spread? A convenient measure of spreading is the root-mean-square displacement $\langle x^2(n)\rangle^{1/2}$. Here we average the square of the displacement rather than the displacement itself. Since the square of a negative number is positive, the result must be finite; it cannot be zero. To find $\langle x^2(n)\rangle$, we write $x_i(n)$ in terms of $x_i(n-1)$, as in Eq. 1.3, and take the square:

$$x_i^2(n) = x_i^2(n-1) \pm 2\delta x_i(n-1) + \delta^2. \quad (1.6)$$

Then we compute the mean,

$$\langle x^2(n)\rangle = \frac{1}{N}\sum_{i=1}^{N} x_i^2(n), \quad (1.7)$$

which is

$$\langle x^2(n)\rangle = \frac{1}{N}\sum_{i=1}^{N} [x_i^2(n-1) \pm 2\delta x_i(n-1) + \delta^2]$$

$$= \langle x^2(n-1)\rangle + \delta^2. \quad (1.8)$$

As before, the second term in the brackets averages to zero; its sign is positive for roughly half of the particles, negative for the other half. Since $x_i(0) = 0$ for all particles i, $\langle x^2(0)\rangle = 0$. Thus, $\langle x^2(1)\rangle = \delta^2$, $\langle x^2(2)\rangle = 2\delta^2$, ..., and $\langle x^2(n)\rangle = n\delta^2$. We conclude that the mean-square displacement increases with the step number n, the root-mean-square displacement with the square-root of n. According to rule 1, the particles execute n steps in a time $t = n\tau$; n is proportional to t. It follows that the mean-square displacement is proportional to t, the root-mean-square displacement to the square-root of t. This is the second proposition. The spreading increases as the square-root of the time, as shown in Fig. 1.3.

To see this more explicity, note that $n = t/\tau$, so that

$$\langle x^2(t)\rangle = (t/\tau)\delta^2 = (\delta^2/\tau)t, \quad (1.9)$$

where we write $x(t)$ rather than $x(n)$ to denote the fact that x now is being considered as a function of t. For convenience, we define a diffusion coefficient, $D = \delta^2/2\tau$, in units cm²/sec. The reason for the factor 1/2 will become clear in Chapter 2. This gives us

$$\langle x^2 \rangle = 2Dt \qquad (1.10)$$

and

$$\langle x^2 \rangle^{1/2} = (2Dt)^{1/2}, \qquad (1.11)$$

where, for simplicity, we drop the explicit functional reference (t). The diffusion coefficient, D, characterizes the migration of particles of a given kind in a given medium at a given temperature. In general, it depends on the size of the particle, the structure of the medium, and the absolute temperature. For a small molecule in water at room temperature, $D \simeq 10^{-5}$ cm²/sec.

A particle with a diffusion coefficient of this order of magnitude diffuses a distance $x = 10^{-4}$ cm (the width of a bacterium) in a time $t \simeq x^2/2D = 5 \times 10^{-4}$ sec, or about half a millisecond. It diffuses a distance $x = 1$ cm (the width of a test tube) in a time $t \simeq x^2/2D = 5 \times 10^4$ sec, or about 14 hours. The difference is dramatic. In order for a particle to wander twice as far, it takes 4 times as long. In order for it to wander 10 times as far, it takes 100 times as long. Therefore, there is no such thing as a diffusion velocity; displacement is not proportional to time but rather to the square-root of the time. What happens if we try to define a diffusion velocity by dividing the root-mean-square displacement by the time? The result is an explicit function of the time. Dividing both sides of Eq. 1.11 by t, we find

$$\frac{\langle x^2 \rangle^{1/2}}{t} = \left(\frac{2D}{t}\right)^{1/2}. \qquad (1.12)$$

Thus, the shorter the period of observation, t, the larger the apparent velocity. For values of t smaller than τ, the apparent velocity is larger than $\delta/\tau = v_x$, the instantaneous velocity of the particle. This is an absurd result. In Chapter 2 we will speak of adsorption rates or diffusion currents. These expressions refer to the number of particles that are adsorbed at, or cross, a given boundary in unit time. They are bulk properties of an ensemble of particles, proportional to their number. They are not rates that tell us how long it takes a particle, by diffusion, to go from here to there. This time depends on the square of the distance, as defined by Eq. 1.10. When next you come across the expression "diffusion rate," think twice! This phrase is ambiguous, at best, and often used incorrectly.

Two- and three-dimensional random walks

Rules 1 to 3 apply in each dimension. In addition, we assert that motions in the x, y, and z directions are statistically independent. If $\langle x^2 \rangle = 2Dt$, then $\langle y^2 \rangle = 2Dt$ and $\langle z^2 \rangle = 2Dt$. In two dimensions, the square of the distance from the origin to the point (x,y) is $r^2 = x^2 + y^2$; therefore,

$$\langle r^2 \rangle = 4Dt. \tag{1.13}$$

In three dimensions, $r^2 = x^2 + y^2 + z^2$, and

$$\langle r^2 \rangle = 6Dt. \tag{1.14}$$

A computer simulation of a two-dimensional random walk is shown in Fig. 1.4. Steps in the x and y directions were made at the same times, so the particle always moved diagonally. The simulation makes graphic a remarkable feature of the random walk, discussed further in Chapter 3. Since explorations over short distances can be made in much shorter times than explorations over long

12—Diffusion: Microscopic Theory

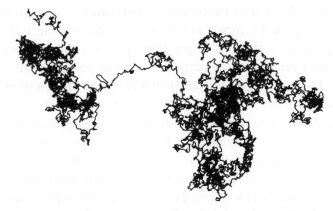

Fig. 1.4. An x, y plot of a two-dimensional random walk of $n = 18{,}050$ steps. The computer pen started at the upper left corner of the track and worked its way to the upper right edge of the track. It repeatedly traversed regions that are completely black. It moved, as the crow flies, 196 step lengths. The expected root-mean-square displacement is $(2n)^{1/2} = 190$ step lengths.

distances, the particle tends to explore a given region of space rather thoroughly. It tends to return to the same point many times before finally wandering away. When it does wander away, it chooses new regions to explore blindly. A particle moving at random has no tendency to move toward regions of space that it has not occupied before; it has absolutely no inkling of the past. Its track does not fill up the space uniformly.

The binomial distribution

We have learned so far that particles undergoing free diffusion have a zero mean displacement and a root-mean-square displacement that is proportional to the square-root of the time. What else can we say about the shape of the distribution of particles? To find out, we have to work out the probabilities that the particles step different distances to the right or to the left. While doing

Diffusion: Microscopic Theory—13

so, it is convenient to generalize the one-dimensional random walk and suppose that a particle steps to the right with a probability p and to the left with a probability q. Since the probability of stepping one way or the other is 1, $q = 1 - p$. The probability that such a particle steps exactly k times to the right in n trials is given by the binomial distribution

$$P(k;n,p) = \frac{n!}{k!(n-k)!} p^k q^{n-k}. \tag{1.15}$$

This equation is derived in Appendix A; see Eqs. A.17, A.18. The displacement of the particle in n trials, $x(n)$, is equal to the number of steps to the right less the number of steps to the left times the step length, δ:

$$x(n) = [k - (n-k)]\delta = (2k - n)\delta. \tag{1.16}$$

Since we know the distribution of k, we know the distribution of x. The two distributions have the same shapes. The probability machine shown in Fig. A.3 converts one into the other.

The mean displacement of the particle is

$$\langle x(n) \rangle = (2\langle k \rangle - n)\delta, \tag{1.17}$$

where

$$\langle k \rangle = np; \tag{1.18}$$

see Eq. A.22. The mean-square displacement is

$$\langle x^2(n) \rangle = \langle [(2k - n)\delta]^2 \rangle$$
$$= (4\langle k^2 \rangle - 4\langle k \rangle n + n^2)\delta^2, \tag{1.19}$$

where

$$\langle k^2 \rangle = (np)^2 + npq; \tag{1.20}$$

see Eq. A.23. For the case $p = q = 1/2$, Eqs. 1.17 and 1.19 yield $\langle x(n) \rangle = 0$ and $\langle x^2(n) \rangle = n\delta^2$, as expected.

The Gaussian distribution

A small particle, such as lysozyme, steps an enormous number of times every second. Given the instantaneous velocity estimated from Eq. 1.2, $v_x = \delta/\tau \simeq 10^3$ cm/sec, and a diffusion coefficient, $D = \delta^2/2\tau \simeq 10^{-6}$ cm^2/sec, we can compute the step length, δ, and the step rate, $1/\tau$. The step length is $2D/v_x \simeq (10^{-6}$ cm^2/sec$)/(10^3$ cm/sec$) = 10^{-9}$ cm, and the step rate is $v_x/\delta \simeq (10^3$ cm/sec$)/(10^{-9}$ cm$) = 10^{12}$ sec^{-1}. Of these $n = 10^{12}$ steps taken each second, $np = 0.5 \times 10^{12}$ are taken to the right. The standard deviation in this number is $(npq)^{1/2} = 0.5 \times 10^6$; see Eq. A.25. So, to a precision of about a part in a million, half of the steps taken each second are made to the right and half to the left. What happens to the distribution of x in this limit? As stated in Appendix A, when n and np are both very large, the binomial distribution, $P(k;n,p)$, is equivalent to

$$P(k)dk = \frac{1}{(2\pi\sigma^2)^{1/2}} e^{-(k-\mu)^2/2\sigma^2} dk, \qquad (1.21)$$

where $P(k)dk$ is the probability of finding a value of k between $k + dk$, $\mu = \langle k \rangle = np$, and $\sigma^2 = npq$; see Eq. A.27. This is the Gaussian or normal distribution. By substituting $x = (2k - n)\delta$, $dx = 2\delta dk$, $p = q = 1/2$, $t = n/\tau$, and $D = \delta^2/2\tau$, we obtain

$$P(x)dx = \frac{1}{(4\pi Dt)^{1/2}} e^{-x^2/4Dt} dx, \qquad (1.22)$$

where $P(x)dx$ is the probability of finding a particle between x and $x + dx$. This is the function plotted in Fig. 1.3. The variance of this distribution is $\sigma_x^2 = 2Dt$; its standard deviation is $\sigma_x = (2Dt)^{1/2}$.

The Gaussian or normal distribution is the distribution encountered most frequently in discussions of propagation of errors. It is tabulated, for example, in the *Hand-*

book of *Chemistry and Physics*, as the "normal curve of error"; see Fig. A.5. About 68% of the area of the curve is within one standard deviation of the origin. Thus, if the root-mean-square displacement of the particles is $(2Dt)^{1/2}$, the chances are 0.32 that a particle has wandered that far or farther. The chances are 0.045 that it has wandered twice as far or farther and 0.0026 that it has wandered three times as far or farther. These numbers are the areas under the curve for $|x| \geq \sigma_x$, $2\sigma_x$, and $3\sigma_x$, respectively.

Visualizing the Gaussian distribution: It is instructive to generate the distributions shown in Fig. 1.3 experimentally. This can be done by layering aqueous solutions of a dye, such as fluorescein or methylene blue, into water. For a first try, layer the dye at the center of a vertical column of water in a graduated cylinder. The dye promptly sinks to the bottom! It does so because it has a higher specific gravity than the surrounding medium. For a second try, match the specific gravity of the medium to the dye by adding sucrose to the water. Now the dye drifts about and becomes uniformly dispersed in a matter of minutes or hours. It does so because there is nothing to stabilize the system against convective flow. Any variation in temperature that increases the specific gravity of regions of the fluid that are higher in the column relative to those that are lower drives this flow. For a final try, layer the dye into a column of water containing more sucrose at the bottom than at the top, i.e., into a sucrose density gradient; a 0-to-2% w/v solution will do. Match the specific gravity of the solution of the dye to that at the midpoint of the gradient and layer it there. Now, patterns of the sort shown in Fig. 1.3 will evolve over a period of many days. The diffusion coefficients of fluorescein, methylene blue, and sucrose are all about

5×10^{-6} cm²/sec. A sucrose gradient $x = 10$ cm high will survive for a period of time of order $t = x^2/2D = 10^7$ sec, or about 4 months. The dye will generate a Gaussian distribution with a standard deviation $\sigma_x = 2.5$ cm in a time $t = \sigma_x^2/2D \simeq 6 \times 10^5$ sec, or in about 1 week. Try it!

It is evident from this experiment that diffusive transport takes a long time when distances are large. Here is another example: The diffusion coefficient of a small molecule in air is about 10^{-1} cm²/sec. If one relied on diffusion to carry molecules of perfume across a crowded room, delays of the order of 1 month would be required. Evidently, the makers of scent owe their livelihood to close encounters, wind, and/or convective flow.

Chapter 2

Diffusion: Macroscopic Theory

Fick's equations

Most discussions of diffusion start with Fick's equations, differential equations that describe the spatial and temporal variation of nonuniform distributions of particles. I find it more illuminating to derive these equations from the model of the random walk. Suppose we know the number of particles at each point along the x axis at time t, as shown in Fig. 2.1. How many particles will move across unit area in unit time from the point x to the point $x + \delta$? What is the net flux in the x direction, J_x? At time $t + \tau$, i.e., after the next step, half the particles at x will have stepped across the dashed line from left to right, and half the particles at $x + \delta$ will have stepped across the dashed line from right to left. The net number crossing to the right will be

$$-\frac{1}{2}[N(x + \delta) - N(x)].$$

To obtain the net flux, we divide by the area normal to the

Fig. 2.1. At time t, there are $N(x)$ particles at position x, $N(x + \delta)$ particles at position $x + \delta$. At time $t + \tau$, half of each set will have stepped to the right and half to the left.

18—Diffusion: Macroscopic Theory

x axis, A, and by the time interval, τ,

$$J_x = -\frac{1}{2}[N(x+\delta) - N(x)]/A\tau.$$

Multiplying by δ^2/δ^2 and rearranging, we obtain

$$J_x = -\frac{\delta^2}{2\tau}\frac{1}{\delta}\left[\frac{N(x+\delta)}{A\delta} - \frac{N(x)}{A\delta}\right].$$

The quantity $\delta^2/2\tau$ is the diffusion coefficient, D. $N(x+\delta)/A\delta$ is the number of particles per unit volume at the point $x+\delta$, i.e., the concentration $C(x+\delta)$. $N(x)/A\delta$ is the concentration $C(x)$. Therefore,

$$J_x = -D\frac{1}{\delta}[C(x+\delta) - C(x)].$$

But δ is very small. In the limit $\delta \to 0$, by the definition of a partial derivative, as explained in Appendix B, we obtain

$$J_x = -D\frac{\partial C}{\partial x}. \qquad (2.1)$$

This is Fick's first equation. It states that the net flux (at x and t) is proportional to the slope of the concentration function (at x and t); the constant of proportionality is $-D$. If the particles are uniformly distributed, the slope is 0, i.e., $\partial C/\partial x = 0$, and $J_x = 0$. If J_x is 0, the distribution will not change with time; the system is at equilibrium. If the slope is constant, i.e., if $\partial C/\partial x$ is constant, J_x is constant. This occurs when C is a linear function of x, as shown in Fig. 2.2. In practice, a gradient of this kind can be maintained only if there is a source for particles at one point and an adsorber for particles at another, e.g., in a pipe connecting reservoirs held at fixed concentrations C_1 and C_2.

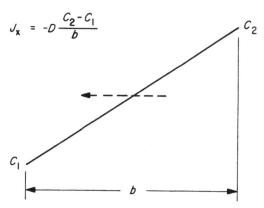

Fig. 2.2. The flux due to a linear concentration gradient $(C_2 - C_1)/b$. There is a net movement of particles from right to left, solely because there are more particles at the right than at the left.

When we derived Eq. 1.10, we defined $D = \delta^2/2\tau$. The reason for the $1/2$ is now clear; it makes Fick's first equation more tidy. Note that if C is expressed in particles/cm^3, J_x is in particles/cm^{2}sec. If C is expressed in moles/cm^3, J_x is in moles/cm^{2}sec.

Fick's second equation follows from the first, provided that the total number of particles is conserved, i.e., that the particles are neither created nor destroyed. Consider the box shown in Fig. 2.3. In a period of time τ, $J_x(x)A\tau$ particles will enter from the left and $J_x(x + \delta)A\tau$ particles will leave from the right. The volume of the box is $A\delta$. If particles are neither created nor destroyed, the number of particles per unit volume in the box must increase at the rate

$$\frac{1}{\tau}[C(t + \tau) - C(t)] = -\frac{1}{\tau}[J_x(x + \delta) - J_x(x)]A\tau/A\delta$$

$$= -\frac{1}{\delta}[J_x(x + \delta) - J_x(x)].$$

20—Diffusion: Macroscopic Theory

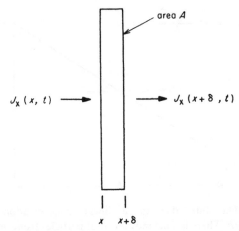

Fig. 2.3. Fluxes through the faces of a thin box extending from position x to position $x + \delta$. The area of each face is A. The faces are normal to the x axis.

In the limit $\tau \to 0$ and $\delta \to 0$, this means that

$$\frac{\partial C}{\partial t} = -\frac{\partial J_x}{\partial x}, \qquad (2.2)$$

or, given Eq. 2.1, that

$$\frac{\partial C}{\partial t} = D\frac{\partial^2 C}{\partial x^2}. \qquad (2.3)$$

This is Fick's second equation. It states that the time rate of change in concentration (at x and t) is proportional to the curvature of the concentration function (at x and t); the constant of proportionality is D. If the slope is constant, $\partial^2 C/\partial x^2 = 0$, and the concentration is stationary: just as many particles diffuse in from the region of higher concentration as diffuse out to the region of lower concentration. Eq. 2.3 tells us how a nonuniform distribution of particles will redistribute itself in time. If we know the initial distribution and other boundary conditions, we can figure out all later distributions.

In three dimensions we have $J_x = -D\,\partial C/\partial x$, $J_y = -D\,\partial C/\partial y$, and $J_z = -D\,\partial C/\partial z$. These are components of a flux vector,

$$\mathbf{J} = -D\,\text{grad}\,C. \tag{2.4}$$

The concentration changes with time as

$$\frac{\partial C}{\partial t} = D\nabla^2 C, \tag{2.5}$$

where ∇^2 is the three-dimensional Laplacian, $\partial^2/\partial x^2 + \partial^2/\partial y^2 + \partial^2/\partial z^2$.

If the problem is spherically symmetric, the flux is radial,

$$J_r = -D\,\partial C/\partial r, \tag{2.6}$$

and

$$\frac{\partial C}{\partial t} = D\frac{1}{r^2}\frac{\partial}{\partial r}\left(r^2\frac{\partial C}{\partial r}\right). \tag{2.7}$$

Time-dependent solutions to Fick's equations

One way to find solutions to Fick's equations is to look them up! An excellent source is Carslaw and Jaeger (1959), a book dealing with the conduction of heat in solids. The heat equation has the same form as the diffusion equation. In the notation of Carslaw and Jaeger,

$$\frac{\partial v}{\partial t} = \kappa \nabla^2 v,$$

where v is the temperature and κ is the thermal diffusivity. So, take their results and read C for v and D for κ. Sources that do not require such translation include Crank (1975) and Jost (1960). But this strategy requires luck. If you happen to find a discussion of just the problem that you are trying to solve, well and good. If not, you will soon be lost in a morass of complex equations. Here are some "trivial" examples.

Diffusion from a micropipette: A micropipette filled with an aqueous solution of a green fluorescent dye is inserted into a large body of water. At time $t = 0$, particles of the dye are injected into the water at the rate i per sec for an infinitesimal period of time dt. The total number of particles injected is $N = idt$. With these boundary conditions, Eq. 2.7 has the solution

$$C(r,t) = \frac{N}{(4\pi Dt)^{3/2}} e^{-r^2/4Dt}. \qquad (2.8)$$

This is a three-dimensional Gaussian distribution; compare Eq. 1.22. Looking through a microscope, one sees the sudden appearance of a green spot that spreads rapidly outward and fades away. The concentration remains highest at the tip of the pipette, but it decreases there as the three-halves power of the time. An observer at radius r sees a wave that peaks at $t = r^2/6D$ at a concentration $C = 0.0736 \, N/r^3$. He finds that the concentration rises most rapidly at time $t = r^2/16.325D$ at a rate $\partial C/\partial t = 1.054 \, ND/r^5$.

The wave due to a pulse of length t_0 can be found by integrating Eq. 2.8 with respect to time. For $t \leq t_0$,

$$C(r,t) = \frac{i}{4\pi Dr} \operatorname{erfc} \frac{r}{(4Dt)^{1/2}}, \qquad (2.9)$$

where erfc x is the error function complement, $1 - \operatorname{erf} x$, and erf x is the error function, defined by the integral

$$\operatorname{erf} x = \frac{2}{\sqrt{\pi}} \int_0^x e^{-u^2} du. \qquad (2.10)$$

This function is tabulated, just like $\sin x$ or $\cos x$; see, for example, Chapter 7 of Abramowitz and Stegun (1972) or Appendix II of Carslaw and Jaeger (1959). Note that erf $0 = 0$, erf $\infty = 1$, and erf $(-x) = -$ erf x. If the pulse is long enough, the concentration approaches the steady-state value

$$C(r,t) = \frac{i}{4\pi Dr}. \quad (2.11)$$

For $t > t_0$,

$$C(r,t) = \frac{i}{4\pi Dr}\left\{\text{erfc}\,\frac{r}{(4Dt)^{1/2}} - \text{erfc}\,\frac{r}{[4D(t-t_0)]^{1/2}}\right\}. \quad (2.12)$$

Figure 2.4 shows the concentration observed at $r = 10^{-3}$ cm for a pulse of constant injection rate i and length $t_0 = 1$ sec for particles of diffusion coefficient $D = 10^{-5}$, 3×10^{-6}, 10^{-6}, 3×10^{-7}, 10^{-7}, and 3×10^{-8} cm²/sec. If the diffusion coefficient is large, the particles diffuse beyond the observer while the pulse is still on, and the peak concentration is given by Eq. 2.11. If the diffusion coefficient is small, the events occur on a time scale that is long compared with the length of the pulse, and the peak concentration approaches $C = 0.0736\ it_0/r^3$, as required by Eq. 2.8. For other examples of diffusion from constrictions, see Jaeger (1965).

Diffusion in a pipe: In principle, one could measure diffusion coefficients with experiments of the kind illustrated in Fig. 2.4, but in practice it is easier to work in one dimension and to follow the spread of a narrow band of particles, as shown in Fig. 1.3. Alternatively, one can expose a column of solution at concentration C_0 to one at concentration 0 and watch the migration of particles from one to the other. In this case, the initial conditions are

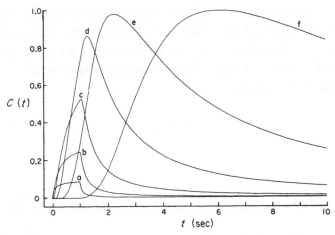

Fig. 2.4. Concentration (in arbitrary units) as a function of time at a distance $r = 10^{-3}$ cm from a point-source in an infinite medium emitting particles at a constant rate from $t = 0$ to $t_0 = 1$ sec, for particles with diffusion coefficient, D (in cm^2/sec): (a) 10^{-5}, (b) 3×10^{-6}, (c) 10^{-6}, (d) 3×10^{-7}, (e) 10^{-7}, and (f) 3×10^{-8}. See Eqs. 2.9 and 2.12.

$C = C_0$ for $x > 0$ and $C = 0$ for $x < 0$, and Eq. 2.3 has the solution

$$C(x,t) = \frac{C_0}{2}\left[1 + \mathrm{erf}\,\frac{x}{(4Dt)^{1/2}}\right]. \qquad (2.13)$$

Since erf $0 = 0$, Eq. 2.13 implies that the concentration of particles at $x = 0$ falls abruptly to $C_0/2$ and remains at that value. This behavior is shown in Fig. 2.5. By taking derivatives of $C(x,t)$ with respect to x or t, we obtain

$$\frac{\partial C}{\partial x} = \frac{C_0}{(4\pi Dt)^{1/2}}\,e^{-x^2/4Dt} \qquad (2.14)$$

and

$$\frac{\partial C}{\partial t} = -\frac{x}{2t}\frac{\partial C}{\partial x}. \qquad (2.15)$$

Equation 2.14 is similar to Eq. 1.22. Thus, D can be mea-

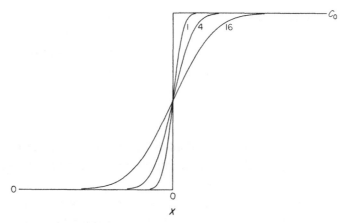

Fig. 2.5. Concentration as a function of position at times $t = 0, 1, 4,$ and 16 for diffusion from a column of liquid initially containing particles at a concentration C_0 (right) into a column of liquid initially devoid of particles (left). The horizontal axis in this figure is drawn on the same scale as that of Fig. 1.3, which shows $\partial C/\partial x$. See Eqs. 2.13–2.15.

sured if one measures C as a function of space and/or time. A numerical solution to a similar problem is given in Appendix B, Fig. B.3.

Steady-state solutions to Fick's equations

If sources and adsorbers are present, the final distribution of particles will not be uniform; instead, the concentration will approach a steady-state value that is higher near sources, lower near adsorbers. In this limit, $\partial C/\partial t = 0$, and Eq. 2.5 reduces to

$$\nabla^2 C = 0. \tag{2.16}$$

For problems with spherical symmetry, Eq. 2.7,

$$\frac{1}{r^2}\frac{d}{dr}\left(r^2 \frac{dC}{dr}\right) = 0. \tag{2.17}$$

We already have seen the steady-state solutions for diffusion in one dimension from a plane at concentration C_2 to a plane at concentration C_1, Fig. 2.2, and for diffusion in three dimensions from a continuous point-source in an infinite medium, Eq. 2.11. Here are some other examples.

Diffusion to a spherical adsorber: Consider a spherical adsorber of radius a in an infinite medium, as shown in Fig. 2.6. Every particle reaching the surface of the sphere is gobbled up, so the concentration at $r = a$ is 0. The concentration at $r = \infty$ is C_0. With these boundary conditions, Eq. 2.17 has the solution

$$C(r) = C_0\left(1 - \frac{a}{r}\right). \tag{2.18}$$

The flux, Eq. 2.6, is

$$J_r(r) = -DC_0 \frac{a}{r^2}. \tag{2.19}$$

The net migration of molecules is radially inward, as shown by the dashed arrows in Fig. 2.6. The particles are

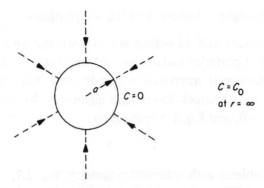

Fig. 2.6. A spherical adsorber of radius a in an infinite medium containing particles at an initial concentration C_0. The dashed arrows are lines of flux.

adsorbed by the sphere at a rate equal to the area, $4\pi a^2$, times the inward flux, $-J_r(a)$:

$$I = 4\pi DaC_0. \qquad (2.20)$$

If C_0 is expressed in particles/cm^3, I is in particles/sec. We will refer to this adsorption rate, I, as a diffusion current. Note that this current is proportional not to the area of the sphere but to its radius. As the radius, a, increases, the area increases as a^2, but the concentration gradient, to which the flux is proportional, decreases as $1/a$.

Diffusion to a disk-like adsorber: Next, consider a disk-shaped adsorber of radius s in a semi-infinite medium, as shown in Fig. 2.7. Every particle reaching the surface of the disk is gobbled up, so the concentration at the disk is 0. The concentration at $x = \infty$ is C_0. This problem is cylindrically symmetric rather than spherically sym-

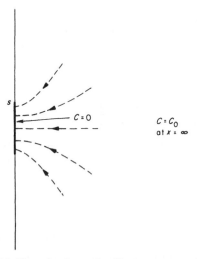

Fig. 2.7. A disk-like adsorber of radius s on one edge of a semi-infinite medium, $x \geq 0$, containing particles at an initial concentration C_0. The dashed arrows are lines of flux.

metric, so the mathematics is not so easy; see, for example, p. 42 of Crank (1975). But the answer turns out to be simple. The diffusion current is

$$I = 4DsC_0. \qquad (2.21)$$

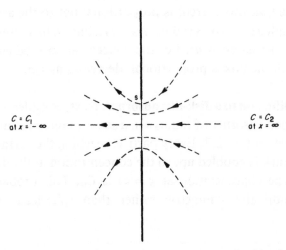

Fig. 2.8. A circular aperture of radius s in a nonadsorbing barrier separating two semi-infinite media, $x < 0$ and $x \geq 0$, containing particles at initial concentrations $C_2 > C_1$. The dashed arrows are lines of flux.

An analogous problem, illustrated in Fig. 2.8, involves diffusion through a circular aperture of radius s in a nonadsorbing barrier separating two semi-infinite media. The concentration at $x = -\infty$ is C_1 and the concentration at $x = \infty$ is C_2. The current through the aperture is

$$I_{2,1} = 2Ds(C_2 - C_1). \qquad (2.22)$$

These currents are proportional not to the area of the disk (or aperture) but to its radius. As the radius, s, increases, the area increases as s^2, but the concentration gradient, to which the flux is proportional, decreases as $1/s$.

Diffusion to an ellipsoidal adsorber: Next, consider a cigar-shaped adsorber, an ellipsoid of revolution with semi-axes $a > b = c$. The concentration at the surface of the ellipsoid is 0, and the concentration at $r = \infty$ is C_0. If the ellipsoid is relatively thin, in particular, if $a^2 \gg b^2$, the diffusion current is

$$I = 4\pi DaC_0/\ln(2a/b), \qquad (2.23)$$

where ln denotes the natural logarithm. This current is smaller than that found for a sphere of radius a given in Eq. 2.20 by the factor $\ln(2a/b)$. This factor is not as large as one might expect. For example, if $b = 10^{-2}a$, $\ln(2a/b) = 5.3$; if $b = 10^{-4}a$, $\ln(2a/b) = 9.9$. Once again, the current is proportional not to the area of the adsorber but roughly to its length.

Appeal to an electrical analogue: The time-independent diffusion equation, Eq. 2.16, is analogous to Laplace's equation for the electrostatic potential in charge-free space. As explained elsewhere (Berg and Purcell, 1977), this implies that the diffusion current to an isolated adsorber of any size and shape can be written as

$$I = 4\pi DcC_0, \qquad (2.24)$$

where c is the electrical capacitance (in cgs units of centimeters) of an isolated conductor of that size and shape. Since the electrical capacitances of a variety of conductors have been worked out, Eq. 2.24 can save some labor. For example, Eq. 2.23 was obtained from an expression for the electrical capacitance of a conducting ellipsoid [by using formula 195.02 of Dwight, 1961, to evaluate the integral 5.02 (4) of Smythe, 1950, and converting from mks to cgs units by multiplying by $1/4\pi\epsilon$]. Smythe used the same integral in another limit to compute the capacitance of a conducting disk, which led us to Eq. 2.21. We could have derived Eq. 2.20 in a similar fashion, but it was more instructive, given the spherical symmetry, to solve the diffusion equation directly.

Diffusion to N disk-like adsorbers on the surface of a sphere: Given that diffusion currents to spheres, disks, and prolate ellipsoids of similar size are roughly equal, it seems likely that diffusion currents to other adsorbers of similar size should be roughly equal. This turns out to be true, even for nonadsorbing objects sprinkled with small adsorptive patches. Suppose N disk-like adsorbers, each of radius s, are distributed over the surface of an otherwise impenetrable nonadsorbing sphere of radius $a \gg s$, as shown in Fig. 2.9. The concentration at $r = \infty$ is C_0.

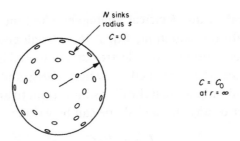

Fig. 2.9. An impenetrable nonadsorbing sphere of radius a covered with N disk-like adsorbers, each of radius s, in an infinite medium containing particles at an initial concentration C_0.

This is a reasonable model for N chemoreceptors or N transport proteins on the surface of a cell. How does the total adsorption rate vary with N? If N is very small, two adsorbers should do twice as well as one, so the rate should increase as $4DNsC_0$, Eq. 2.21. But when N is very large, almost the entire surface of the sphere is adsorbing, and the rate should approach $4\pi DaC_0$, Eq. 2.20. What happens in between? In this regime, the distance between

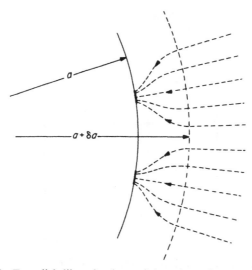

Fig. 2.10. Two disk-like adsorbers of the sphere shown in Fig. 2.9. The dashed arrows are lines of flux. These lines become radial at a distance δa from the surface of the sphere, a distance roughly equal to the distance between adjacent adsorbers.

receptors is large compared with their radius but small compared with the radius of the cell, as shown in Fig. 2.10. The lines of flux are radial for $r > a + \delta a$ (as in Fig. 2.6) but converge on the adsorbers for $a < r < a + \delta a$ (as in Fig. 2.7). Evidently, the concentration at $r = a + \delta a$ is constant at some intermediate value between 0 and C_0.

This problem is formally equivalent to one in electricity in which current flows through a medium of finite resistivity to N conductive patches on an insulating sphere, with the medium a large distance away at potential V_0 and the patches at potential 0. The concentration, C, is an analogue of the potential, V. In the electrical case, we have Ohm's law, which states that the current through a resistor is equal to the potential drop across its terminals divided by its resistance. For steady-state diffusion, we have $I = C/R$, where I is the diffusion current, C is the concentration difference, and R is the diffusion resistance.

32—Diffusion: Macroscopic Theory

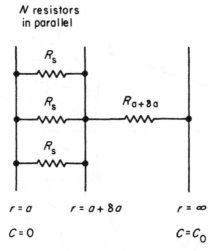

Fig. 2.11. An electrical model for the problem of N adsorbers of radius s on the surface of a sphere of radius a, the problem illustrated in Figs. 2.9 and 2.10.

By appealing to this relation, we note that the diffusion resistance for the adsorbing sphere, Eq. 2.20, is $R_a = 1/4\pi Da$, and that the diffusion resistance for the disk-like adsorber, Eq. 2.20, is $R_s = 1/4Ds$. These resistances are shown as discrete elements in Fig. 2.11. The total resistance of this circuit is $R = R_{a+\delta a} + R_s/N = 1/[4\pi D(a + \delta a)] + 1/4DNs$. Since $\delta a \ll a$, $R \simeq 1/4\pi Da + 1/4DNs = (1/4\pi Da)(1 + \pi a/Ns) = R_a(1 + \pi a/Ns)$. We conclude that the diffusion resistance for a sphere covered with N disk-like adsorbers is larger than the diffusion resistance for the completely adsorbing sphere by a factor $1 + \pi a/Ns$. The diffusion current is smaller by the same factor:

$$\frac{I}{I_0} = \frac{1}{1 + \pi a/Ns}, \tag{2.25}$$

where $I_0 = 4\pi DaC_0$, Eq. 2.20. I is plotted as a function of N in Fig. 2.12. If N is small, the rate increases as $4DNsC_0$. If N is large, the rate approaches $4\pi DaC_0$. This is the asymptotic behavior that we predicted.

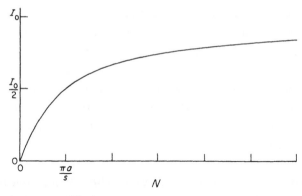

Fig. 2.12. The diffusion current, or rate of adsorption, I, as a function of the number of adsorbers, N, for disk-like adsorbers of radius s on the surface of a sphere of radius a. See Eq. 2.25. I_0 is the diffusion current for the completely adsorbing sphere, Eq. 2.20.

Note that the diffusion current reaches half of its maximum value for $N = \pi a/s$. This number is surprisingly small. Consider a spherical cell of radius $a = 5$ μm equipped with N transport proteins, each with a binding site of radius $s = 10$ Å. This cell can adsorb substrate molecules at half the rate of a cell completely covered by such sites if $N = \pi a/s = 15{,}700$. Only a small fraction of the surface of the cell need be specifically adsorbing, namely $N\pi s^2/4\pi a^2 = 1.6 \times 10^{-4}$. The distance between neighboring sites is about $(4\pi a^2/N)^{1/2} = 0.14$ μm, i.e., about 140 times the site radius. Thus, many hundreds of different transport (or receptor) systems can be accommodated on the surface of the cell, each adsorbing particles of a specific kind with an efficiency approaching that of a cell whose entire surface is dedicated to one such task. For other discussions of this problem, see Berg and Purcell (1977), DeLisi and Wiegel (1981), Shoup and Szabo (1982), and Zwanzig (1990). For an application involving detection of insect pheromones, see Futrelle (1984).

34—Diffusion: Macroscopic Theory

It should be stressed that the electrical analogy used in the derivation of Eq. 2.25 does not extend to time-dependent diffusion; it only applies in the steady-state. The flux is proportional to the concentration gradient, but individual particles are not moving like electrons through a wire; they are moving strictly at random. The same restriction applies to the analogy leading to Eq. 2.24.

Diffusion through N circular apertures in a planar barrier: Consider a system in which two plates a distance b apart are held at concentrations C_1 and C_2, as shown in

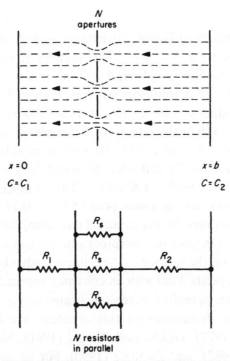

Fig. 2.13. A barrier with $N = nA$ apertures, each of radius s, separating two plates of area A held at concentrations C_1 and C_2 (top), and an electrical model for this system (bottom). See Eq. 2.27.

Fig. 2.2. Let the system have a cross-sectional area A. The diffusion current from one plate to the other is

$$I_{2,1} = DA(C_2 - C_1)/b. \qquad (2.26)$$

Now insert somewhere between the two plates a thin barrier containing N apertures, each of radius $s \ll b$, as shown in Fig. 2.13. What is the diffusion current in the new steady state? The diffusion resistance for one aperture is $R_s = 1/2Ds$, Eq. 2.22. The diffusion resistance of the medium between the plates is $R_1 + R_2 \simeq R_{2,1} = b/DA$, Eq. 2.26. The diffusion resistance of the entire system is $R_1 + R_2 + R_s/N \simeq R_{2,1} + R_s/N = b/DA + 1/2DNs = (b/DA)(1 + 1/2nsb)$, where n is the number of apertures per unit area. It follows that the barrier decreases the diffusion current by a factor

$$\frac{I}{I_{2,1}} = \frac{1}{1 + 1/2nsb}. \qquad (2.27)$$

If this ratio is written as $b/(b + 1/2ns)$, it is evident from the denominator that the effect of the barrier is equivalent to the addition of an extra diffusion path of length $1/2ns$.

The diffusion current reaches half of its maximum value for $n = 1/2sb$. This number is surprisingly small. Consider two cubical cells 10 μm on a side joined on one face. Assume that the membranes composing this face are of negligible thickness and penetrated by N pores, each of radius 50 Å. How many pores are required for the diffusion current between the cells to be half as large as it would be were the barrier not there? Given $N/A = 1/2sb$, with $A = 10^{-6}$ cm^2, $s = 5 \times 10^{-7}$ cm, and $b = 2 \times 10^{-3}$ cm, we find $N = 500$. Only a small fraction of the surface of the barrier need be penetrated, namely $N\pi s^2/A = 3.9 \times 10^{-4}$. The distance between neighboring pores is about $(A/N)^{1/2} = 0.45$ μm, i.e., about 90 times the pore radius.

A similar problem involves the diffusion of gases through the stomata of leaves. In this case, the concentration is clamped at 0 inside the leaf, just to the left of the barrier; so $R_1 = 0$, and the diffusion resistance for each aperture is the same as that for the disk-like adsorber, $R_s = 1/4Ds$. Thus, the stomata add an extra diffusion path of length $1/4ns$, and the diffusion current reaches half of its maximum value for $n = 1/4sb$. This analysis is valid only for the boundary layer near the surface of the leaf, i.e., in still air. For a discussion of this problem, see Chapter 3 of Meidner and Mansfield (1968).

Chapter 3
Diffusion to Capture

In Chapter 2 we compared the steady-state rates of uptake of particles by completely adsorbing objects of various shapes, such as spheres, disks, and ellipsoids. We found that these rates are proportional to the linear dimension of the object rather than to its area, and that the shape is not of crucial importance. For example, the diffusion currents to a sphere of radius a, a two-sided disk of radius a, and an ellipsoid of revolution of length $2a$ and radius $a/10$ fall in the ratios 1 to 0.64 to 0.33. We also found that a reflecting object of a given size and shape sparsely covered with adsorbent patches is nearly as good at sequestering particles as a completely adsorbing object of the same size and shape. From a microscopic point of view, both sets of results reflect the fact that a diffusing particle that finds itself in a given region of space is destined, by that very circumstance, to wander around that region for a time, probing it rather thoroughly before wandering away for good. A particle that finds itself in a spherical space of radius a has a fair chance of blundering into a disk or an ellipsoid inscribed in that space. A particle that bumps into a reflecting object has a fair chance of hitting an adsorbent patch nearby on its surface. This property of the random walk is apparent in the two-dimensional simulation shown in Fig. 1.4. That particle wandered about in some regions at great length but ignored others completely. It is very difficult to get a feel for this mindless ramble from a casual study of the diffusion equation (Eq. 2.5). In this chapter, we try to make these ideas more quantitative by working out some probabilities of capture and mean times to capture.

Probability of capture

Suppose a particle is released near a spherical adsorber of radius a at a point $r = b > a$? What is the probability that the particle will be adsorbed at $r = a$ rather than wander away for good? Naively, one might think that as b increases, this probability would decrease as $1/b^2$, as would be expected if the probability of capture depended on the solid angle subtended by the adsorber at the point of release. In fact, the probability decreases only as $1/b$.

To see this, consider a spherical shell source of radius b between a spherical adsorber of radius a and a spherical shell adsorber of radius c, as shown in Fig. 3.1. The concentration rises from 0 at $r = a$ to a maximum value C_m at $r = b$ and then falls again to 0 at $r = c$. With these

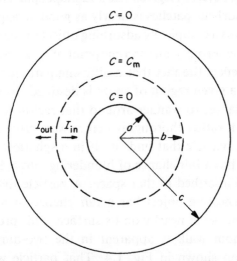

Fig. 3.1. A spherical shell source, radius b, between a spherical adsorber of radius a and a spherical shell adsorber of radius c. Particles released at $r = b$ move inward and are adsorbed at $r = a$ at rate I_{in} or move outward and are adsorbed at $r = c$ at rate I_{out}. Their steady-state concentration rises from 0 at $r = a$ to C_m at $r = b$ and then falls again to 0 at $r = c$. See Eqs. 3.1–3.5.

boundary conditions, Eq. 2.17 has the solution

$$C(r) = \begin{cases} \dfrac{C_m}{1 - a/b}\left(1 - \dfrac{a}{r}\right) & a \leq r \leq b \\ \dfrac{C_m}{c/b - 1}\left(\dfrac{c}{r} - 1\right) & b \leq r \leq c. \end{cases} \quad (3.1)$$

The radial flux, Eq. 2.6, is

$$J_r(r) = \begin{cases} -\dfrac{DC_m}{1 - a/b}\dfrac{a}{r^2} & a \leq r \leq b \\ \dfrac{DC_m}{c/b - 1}\dfrac{c}{r^2} & b \leq r \leq c. \end{cases} \quad (3.2)$$

Thus, the diffusion current from the spherical shell source to the inner adsorber is

$$I_{in} = 4\pi DC_m \frac{a}{1 - a/b}, \quad (3.3)$$

and the diffusion current from the spherical shell source to the outer adsorber is

$$I_{out} = 4\pi DC_m \frac{c}{c/b - 1}. \quad (3.4)$$

The ratio

$$\frac{I_{in}}{I_{in} + I_{out}} = \frac{a(c - b)}{b(c - a)} \quad (3.5)$$

is the probability that a particle released at $r = b$ will be adsorbed at $r = a$. In the limit $c \to \infty$, this probability is just a/b. This is the probability of capture for the sphere of radius a immersed in an infinite medium. As b increases, this probability decreases as $1/b$, as predicted.

An individual particle has no way of knowing that an adsorber is present unless it blunders into it. Thus, Eq. 3.5 gives the probability that a particle released at $r = b$ reaches $r = a$ before diffusing as far as $r = c$, even when the adsorbers are not there. Suppose a particle is released at $r = b$ near the surface of a reflecting sphere of radius a. How many times, on the average, does it reach the surface of the sphere and return to $r = b$ before wandering away for good? The probability that the particle visits the sphere at least once before wandering away for good is $p = a/b$. The probability that it visits the sphere once, returns to $r = b$, and then wanders away for good is $p(1 - p)$. The probability that the particle makes this round trip twice and then wanders away for good is $p^2(1 - p)$. The probability that it does so n times is $p^n(1 - p)$. Therefore, the mean number of round trips is

$$\langle n \rangle = \sum_{n=0}^{\infty} n p^n (1 - p). \tag{3.6}$$

By factoring out $p(1 - p)$ and noting that $1 + 2p + 3p^2 + \cdots = (1 - p)^{-2}$, we find

$$\langle n \rangle = p/(1 - p) = a/(b - a). \tag{3.7}$$

It follows that a particle close to the surface of the sphere makes a large number of trips from b to a and back again before wandering away for good; the number increases without limit as b approaches a. In the process, the particle probes the sphere at a large number of points, some far enough apart to allow it to find adsorbent patches, even when these patches cover only a small fraction of the surface. For a derivation of Eq. 2.25 from this point of view, see pp. 196-198 of Berg and Purcell (1977).

What is the probability that a particle released at $r = b$ near the surface of an adsorbing sphere of radius a wanders as far as $r = c$ or farther before being captured? The probability that a particle is adsorbed at $r = a$ after diffusing to an arbitrary radius is a/b. The probability that a particle is adsorbed at $r = a$ without diffusing as far as $r = c$ is given by Eq. 3.5. The probability that a particle is adsorbed at $r = a$ after diffusing as far as $r = c$ or farther is the difference of these two probabilities, namely, $a/b - a(c - b)/b(c - a) = a(b - a)/b(c - a)$. The fraction of particles that do so is this number divided by a/b, i.e., $(b - a)/(c - a)$. Thus, on the average, if 100 particles are released at $r = 2a$, $(a/2a) \times 100 = 50$ will be adsorbed at $r = a$, and 50 will wander away for good. Of the 50 that are adsorbed at $r = a$, $[(2a - a)/(3a - a)] \times 50 = 25$ will have wandered as far as $r = 3a$ or farther before being captured. One will have wandered as far as $r = 51a$ or farther before being captured.

We noted earlier that a particle executing a random walk has no inkling of its past. These calculations emphasize the fact that it also has no inkling of its future. The behavior of an individual particle is not affected by an adsorber unless, by chance, it happens to bump into it.

The one-dimensional case is shown in Fig. 3.2. The diffusion currents are $I_{left} = DAC_m/a$, and $I_{right} = DAC_m/(b - a)$. The probability that a particle released at $x = a$ will be adsorbed at $x = 0$ is

$$\frac{I_{left}}{I_{left} + I_{right}} = \frac{b - a}{b}. \tag{3.8}$$

In the limit $b \to \infty$, this probability is 1. The average number of trips that a particle makes from $x = a$ to a reflecting barrier at $x = 0$ before reaching $x = b$ is $(b - a)/a$. The probability that a particle released at $x = a$ is adsorbed at $x = 0$ after diffusing as far as $x = b$ or farther is

$1 - (1 - a/b) = a/b$. Thus, on the average, if 100 particles are released at $r = a$, 50 will have wandered as far as $x = 2a$ or farther before being captured. One will have wandered as far as $x = 100a$ or farther before being captured. None of the particles will wander away for good.

Mean time to capture

When a particle is released at position $x = a$, as indicated in Fig. 3.2, how long does it take to blunder into an adsorber at $x = 0$ or $x = b$? If this experiment is repeated many times, what is the mean time to capture, $W(a)$? To find out, we return to the formalism of the random walk, release a particle at position x at time $t = 0$, and allow it to step to the right or to the left a distance δ every τ sec. At time τ, the particle will be at position $x + \delta$ with probability $1/2$, or at position $x - \delta$ with probability $1/2$. The mean times to capture from these positions are $W(x + \delta)$

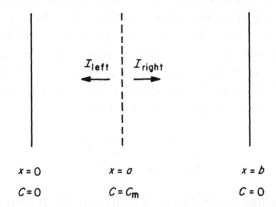

Fig. 3.2. A planar source at $x = a$ between two planar adsorbers at $x = 0$ and $x = b$. The cross-sectional area of the system is A. Particles released at $x = a$ move to the left and are adsorbed at $x = 0$ at rate I_{left} or move to the right and are adsorbed at $x = b$ at rate I_{right}. Their steady-state concentration rises from 0 at $x = 0$ to C_m at $x = a$ and then falls again to 0 at $x = b$. See Eq. 3.8.

and $W(x - \delta)$, respectively. Thus, the expectation value of $W(x)$ is

$$W(x) = \tau + \frac{1}{2}[W(x + \delta) + W(x - \delta)]. \quad (3.9)$$

By subtracting $W(x)$ from both sides and multiplying through by $2/\delta$, we obtain

$$\frac{1}{\delta}[W(x + \delta) - W(x)] - \frac{1}{\delta}[W(x) - W(x - \delta)]$$

$$+ \frac{2\tau}{\delta} = 0.$$

In the limit of very small δ, by the definition of a derivative, we have

$$\left.\frac{dW}{dx}\right|_x - \left.\frac{dW}{dx}\right|_{x-\delta} + \frac{2\tau}{\delta} = 0.$$

Dividing once more by δ, appealing again to the definition of a derivative, and noting that $2\tau/\delta^2 = 1/D$, we obtain

$$\frac{d^2W}{dx^2} + \frac{1}{D} = 0. \quad (3.10)$$

This differential equation can be solved for W given suitable boundary conditions. At an adsorbing boundary, the mean time to capture is 0, so $W = 0$. At a reflecting boundary, the mean time to capture does not vary with x, so $dW/dx = 0$.

If there are adsorbing boundaries at $x = 0$ and $x = b$, as shown in Fig. 3.2, then $W(0) = W(b) = 0$, and Eq. 3.10 has the solution

$$W(x) = \frac{1}{2D}(bx - x^2). \quad (3.11)$$

The mean time to capture a particle released halfway in between, at $x = b/2$, is $b^2/8D$. The mean time to capture a particle released at random anywhere in between $x = 0$ and $x = b$ is given by the average

$$\frac{1}{b}\int_0^b W(x)\,dx, \tag{3.12}$$

which, on substitution of Eq. 3.11, gives $b^2/12D$.

If there is an adsorbing boundary at $x = 0$ and a reflecting boundary at $x = b$, then $W(0) = 0$, $dW/dx = 0$ at $x = b$, and

$$W(x) = \frac{1}{2D}(2bx - x^2). \tag{3.13}$$

The mean time to capture a particle released at $x = b/2$ is now $3b^2/8D$, three times longer than before. Sometimes a particle released at $x = b/2$ wanders from $x = b/2$ to $x = b$ and back again before capture at $x = 0$. Eq. 3.8 tells us that the probability for this is 1/2. Events of this kind raise the mean. The mean time to capture a particle released at random is now $b^2/3D$, four times longer than before.

These results are shown graphically in Fig. 3.3. The average height of each curve is given by the area that it subtends divided by its width, as specified by Eq. 3.12. In the case of two adsorbing boundaries, the largest contributions to the average occur when the particle is released near $x = b/2$; in the case of one adsorbing and one reflecting boundary, the largest contributions occur when the particle is released near $x = b$.

An example of a one-dimensional process of this kind of enormous practical significance is the diffusion of a molecule of repressor along a strand of DNA, in quest of its binding site (the operator). If the repressor, after

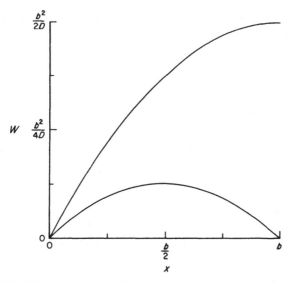

Fig. 3.3. Plots of the mean time to capture for particles released at position x with adsorbing boundaries at $x = 0$ and $x = b$ (lower curve, Eq. 3.11) or with an adsorbing boundary at $x = 0$ and a reflecting boundary at $x = b$ (upper curve, Eq. 3.13). If x is chosen at random, the mean time to capture is $b^2/12D$ or $b^2/3D$, respectively; see Eq. 3.12.

having arrived at a random point on a segment of DNA of length b terminated at one end by the operator, is adsorbed strongly enough that it cannot desorb but loosely enough that it can move along the DNA with a finite diffusion coefficient D', then its mean time to capture is $b^2/3D'$. Given the surprisingly high rate at which particles can be adsorbed by a thread-like object, Eq. 2.23, it is reasonable to suppose that the repressor actually finds its binding site by such a two-stage mechanism. But, as discussed for an analogous problem in two dimensions (pp. 198-200 of Berg and Purcell, 1977), the importance of this mechanism depends on the strength of the nonspecific binding. Unless it is strong, the repressor can do as well or better by staying in three dimensions.

The equation for the mean time to capture in two or three dimensions is

$$\nabla^2 W + \frac{1}{D} = 0, \qquad (3.14)$$

where ∇^2 is the two- or three-dimensional Laplacian. This equation also can be solved by appeal to an electrical analogue; it is Poisson's equation for the potential (W) in a region of uniform charge density ($1/4\pi D$). For some sample solutions, see Appendix B of Berg and Purcell (1977).

Note, finally, that the logic used in the derivation of Eq. 3.10 can be applied as well to the problem posed at the beginning of this chapter, the probability of capture of a particle at an adsorbing boundary. The recursion relation for the probability of capture, $P(x)$, of a particle released at point x, corresponding to the recursion relation for the mean time to capture, Eq. 3.9, is

$$P(x) = \frac{1}{2} [P(x + \delta) + P(x - \delta)]. \qquad (3.15)$$

This leads to the differential equation

$$\frac{d^2 P}{dx^2} = 0. \qquad (3.16)$$

The boundary conditions for capture by an adsorber at $x = 0$ rather than at one at $x = b$ are $P(0) = 1$ and $P(b) = 0$. With these boundary conditions, Eq. 3.16 has the solution

$$P(x) = \frac{b - x}{b}, \qquad (3.17)$$

which, for $x = a$, is the result obtained earlier from computation of diffusion currents, Eq. 3.8.

In two or three dimensions we have

$$\nabla^2 P = 0, \qquad (3.18)$$

where, as before, ∇^2 is the two- or three-dimensional Laplacian. For a particle released near a spherical adsorber of radius a at a point $r = b > a$ the boundary conditions are $P(a) = 1$ and $P(\infty) = 0$. With these boundary conditions, Eq. 3.18 has the solution

$$P(r) = \frac{a}{r}, \qquad (3.19)$$

which, for $r = b$, is the result obtained earlier from computation of diffusion currents, a/b. Equation 3.5 follows from the solution of Eq. 3.18 for a spherical adsorber of radius a inside a spherical shell adsorber of radius c with boundary conditions $P(a) = 1$ and $P(c) = 0$.

Chapter 4

Diffusion with Drift

The theory of diffusion developed in Chapters 1–3 would be more useful if we had a means of estimating the values of diffusion coefficients from first principles, given the sizes and shapes of the particles of interest. It turns out that this can be done if we compute the velocity at which a particle drifts through the medium when exposed to an externally applied force, such as that due to a gravitational, centrifugal, or electrical field. In practice, the velocity at which the particle moves in response to such a field is infinitesimal when compared with the instantaneous root-mean-square velocity given by Eq. 1.2. This means that the particles diffuse much as they would in the absence of the field, but with a small persistent directional bias, as indicated in Fig. 1.1c.

Random walk with drift

Consider a particle of mass m at position x subjected to an externally applied force, F_x, acting in the $+x$ direction, as shown in Fig. 4.1. In accordance with Newton's second law, the force causes the particle to accelerate uni-

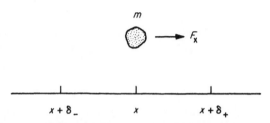

Fig. 4.1. A particle of mass m subjected to an externally applied force F_x while undergoing a one-dimensional random walk.

formly to the right with acceleration $a = F_x/m$. The random walk proceeds as before, according to the rules set down in Chapter 1, with a particle stepping to the right or the left once every τ seconds with an initial velocity $+v_x$ or $-v_x$. A particle starting at position x with an initial velocity $+v_x$ moves in time τ a distance $\delta_+ = v_x\tau + a\tau^2/2$, while a particle starting at position x with an initial velocity $-v_x$ moves in time τ a distance $\delta_- = -v_x\tau + a\tau^2/2$. Since steps to the right and left are equally probable, the average displacement in time τ is $a\tau^2/2$, and the particle drifts to the right with an average velocity

$$v_d = \frac{1}{2} a\tau = \frac{1}{2} \frac{F_x}{m} \tau. \tag{4.1}$$

It is customary to relate the drift velocity to the applied force by a parameter, f, called the frictional drag coefficient:

$$v_d = \frac{F_x}{f}. \tag{4.2}$$

In our model, $f = 2m/\tau$. Multiplying both the numerator and the denominator of this expression by $(\delta/\tau)^2$ and noting that $v_x = \delta/\tau$ and $D = \delta^2/2\tau$, we find $f = mv_x^2/D$. But by Eq. 1.1, $mv_x^2 = kT$; therefore, $f = kT/D$, or

$$D = \frac{kT}{f}. \tag{4.3}$$

This result, known as the Einstein-Smoluchowski relation, turns out to be very general. It does not depend on any assumptions made about the structure of the particle or the details of its motion, a point to which we will return in Chapter 5. Given Eqs. 4.2 and 4.3, we have a procedure for estimating D. First, apply a force F_x, measure v_d, and use Eq. 4.2 to compute f; then, use Eq. 4.3 to compute D.

A reader who knows more physics might be perturbed by our derivation of Eq. 4.3. Real particles do not step in synchrony at a fixed interval, move solely in one dimension, or start each step at a fixed velocity. Step intervals, directions, velocities, and lengths continuously vary as the particle exchanges energy with the molecules of the fluid in which it is suspended. In a more rigorous treatment, one worries about the distributions of these quantities and defines a mean collision time—or for a large particle diffusing in a medium of small particles, a directional correlation time—a mean velocity, and a mean free path. The functional dependence of D and f on these parameters is the same as in our model, but some of the numerical coefficients differ. The final result is the same. The essential point is that a particle is accelerated by the externally applied force; it forgets about this acceleration when it exchanges energy with the molecules of the fluid in which it is suspended, and then it is accelerated once again. As a result, the particle drifts through the medium with a velocity proportional to the externally applied force. The constant of proportionality is D/kT. For further discussion of these points, see Chapter I-43 of Feynman, Leighton, and Sands (1963).

Note, finally, that we could have obtained the same drift velocity, $v_d = (D/kT)F_x$, from a biased random walk, with the step rate, velocity and distance constant, given a probability of stepping in the $+x$ direction $p = 1/2 + F_x\delta/4kT$ and a probability of stepping in the $-x$ direction $q = 1/2 - F_x\delta/4kT$. To see this, use Eqs. 1.17 and 1.18 with $n = t/\tau$, $D = \delta^2/2\tau$. Since the bias is small, $p/q = 1 + F_x\delta/kT$.

Fick's equations for diffusion with drift

If all the particles in a distribution drift in the $+x$ direction at velocity v_d, then the flux at point x must increase by an amount $v_d C(x)$. Thus, Fick's first equation, Eq. 2.1, becomes

$$J_x = -D\frac{\partial C}{\partial x} + v_d C. \qquad (4.4)$$

The derivation of Fick's second equation, Eq. 2.3, proceeds as before, giving

$$\frac{\partial C}{\partial t} = D\frac{\partial^2 C}{\partial x^2} - v_d \frac{\partial C}{\partial x}. \qquad (4.5)$$

We will use this equation in the general proof of the Einstein-Smoluchowski relation given in Chapter 5.

Viscous drag

If a particle is large compared with the molecules composing the medium in which it is suspended, it is possible to use the equations of motion of viscous fluids and calculate the frictional drag coefficient. These equations, called the Navier-Stokes equations, become relatively simple when applied to small things moving slowly, i.e., under conditions in which viscous forces are important but inertial forces are not. Viscous forces arise whenever a fluid is sheared, i.e., whenever the velocities of adjacent regions of fluid differ. Shear is generated, for example, when two parallel plates are moved relative to one another, as shown in Fig. 4.2. In this case, the velocity

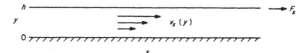

Fig. 4.2. A viscous fluid contained between flat plates at $y = 0$ and $y = h$. The area of each plate is A. The bottom plate is fixed. The top one is propelled to the right by a force in the x direction, F_x. Arrows show the velocity of the fluid relative to the bottom plate at different distances from this plate, y.

profile is linear:

$$v_x(y) = \frac{v_x(h)}{h} y, \qquad (4.6)$$

and the shear, $\partial v_x/\partial y$, does not depend on the position, y. Note that there is no slip at the boundaries; the fluid at the surface of each plate moves at the velocity of that plate. The force, F_x, is balanced by the viscous drag due to the shear:

$$F_x = \eta A \left.\frac{\partial v_x}{\partial y}\right|_{y=h}. \qquad (4.7)$$

This equation provides an operational definition for the coefficient of viscosity, η. The viscous drag is proportional to η, the area of the plate, and the rate of shear of the fluid at the surface of the plate. The units of η can be found from the dimensional equation $[\text{g cm/sec}^2]$ = $[\eta][\text{cm}^2][\text{cm/sec cm}]$, which yields $[\eta] = [\text{g/cm sec}]$, which is called a poise (P). The kinematic viscosity, ν, is η divided by the specific gravity of the medium, ρ; its units are the same as that of the diffusion coefficient, cm²/sec, which is called a stoke. The viscosity of water at 20°C is 0.01 P, or 1 centipoise (cP). The viscosities of air, water, and glycerol are compared in the table below.

Viscosities of various fluids at different temperatures, in g/cm sec (poise)

	Air	Water	Glycerol (dry)
40°C	1.93×10^{-4}	6.53×10^{-3}	2.83
20°C	1.83×10^{-4}	1.00×10^{-2}	14.1
0°C	1.71×10^{-4}	1.79×10^{-2}	120.

The viscosity of air increases slowly with temperature; it is nearly independent of pressure (data not shown). The viscosities of the liquids decrease rapidly with temperature. For most liquids encountered in the laboratory, η is a constant at a given temperature; it does not depend on the rate of shear. However, this is not the case for solutions containing molecules with long unbranched chains, such as methylcellulose; see, for example, Berg and Turner (1979). Indeed, some media are so complex that the viscous drag is not even in the direction of flow. Fortunately, we do not encounter these problems with dilute aqueous solutions of globular particles or with substances of low molecular weight.

Flow through a thin rectangular channel: Here is a sample calculation. Consider a channel of length b bounded by flat plates at $y = 0$ and $y = h$, as shown in Fig. 4.3. If a pressure difference ΔP is applied between the left and right ends of this channel, what is the flow profile $v_x(y)$? We can solve this problem using Eq. 4.7, assuming that the width of the channel, w, is so large compared with its height, h, that edge effects can be neglected. Consider a thin layer of fluid extending from y to $y + dy$. The net force due to the pressure difference at the ends of this layer tending to drive the fluid through the channel is $\Delta Pwdy$. The net drag due to viscous shear at the bottom and top edges of this layer tending to resist such flow is

$$\eta bw[(\partial v_x/\partial y)_y - (\partial v_x/\partial y)_{y+dy}] = -\eta bw(\partial^2 v_x/\partial y^2)dy.$$

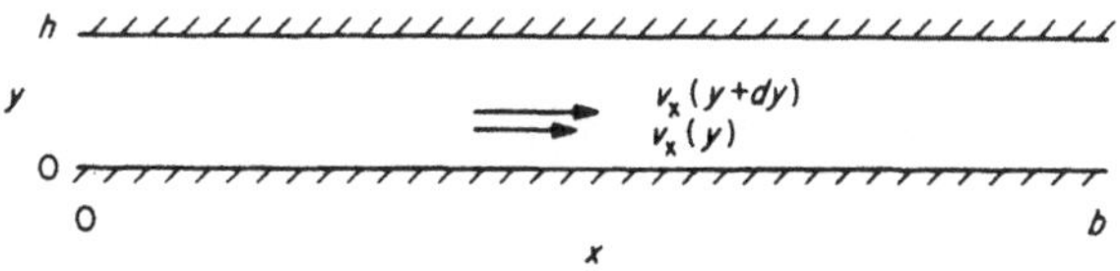

Fig. 4.3. Fluid in a channel between two parallel plates driven from left to right by a pressure gradient $\Delta P/b$. Both plates are fixed. The channel is of height h, length b, and width $w \gg h$.

Thus,

$$\frac{\partial^2 v_x}{\partial y^2} = -\frac{\Delta P}{\eta b}. \tag{4.8}$$

Integrating this equation twice and applying the boundary conditions $v_x(0) = v_x(h) = 0$, we obtain

$$v_x(y) = \frac{\Delta P}{2\eta b}(hy - y^2) = \frac{4v_m}{h^2}(hy - y^2), \tag{4.9}$$

where v_m is the maximum velocity, $\Delta P h^2 / 8\eta b$. The flow profile is parabolic; the maximum velocity occurs in the middle of the channel at $y = h/2$. The volume of fluid passing through the channel per unit time is

$$w \int_0^h v_x(y) dy = \frac{2}{3} w h v_m. \tag{4.10}$$

The average velocity of the fluid is $2v_m/3$.

In its most general form, Eq. 4.8 states that the Laplacian of the velocity is equal to the gradient of the pressure divided by the coefficient of viscosity. This is the basic equation of slow viscous flow; see, for example, Chapter 2 of Landau and Lifshitz (1959).

Flow around a sphere: Viscous flows around small particles are highly regular. Figure 4.4 shows the flow lines around a sphere, radius a, moving to the right through a viscous fluid at constant velocity v_d. The radial and azimuthal components of the velocity of the fluid relative to the sphere are

$$v_r = -v_d \cos\theta \left(1 - \frac{3a}{2r} + \frac{a^3}{2r^3}\right)$$

$$v_\theta = v_d \sin\theta \left(1 - \frac{3a}{4r} - \frac{a^3}{4r^3}\right). \tag{4.11}$$

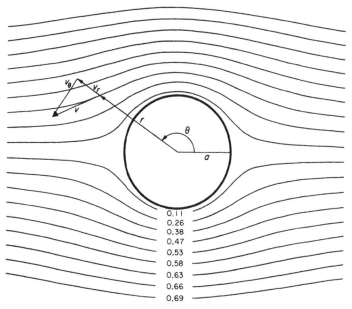

Fig. 4.4. A solid sphere moving at a constant velocity v_d through an incompressible viscous fluid. The fluid moves around the sphere along the flow lines shown. See Eq. 4.11. The numbers on the flow lines at $\theta = -90°$ indicate the magnitudes of v_θ at these points in units of v_d.

Note that both are 0 at the surface of the sphere and that both increase as r increases. The fluid at the surface of the sphere moves with the sphere; the fluid a long distance away does not. The motion of the sphere causes the fluid to shear. Even at the outermost flow line shown in the figure, the fluid is moving 30% as fast as the sphere. The sphere carries fluid with it. Evidently, it must move several diameters before it can shed its local environment.

Stokes' law: A calculation of the net force acting on the sphere yields Stokes' law, which states that the force necessary to drag a sphere of radius a at a velocity v_d through an incompressible, viscous liquid is $6\pi\eta a v_d$.

Given Eqs. 4.2 and 4.3, this implies that

$$f_{sphere} = 6\pi\eta a, \quad (4.12)$$

and

$$D_{sphere} = \frac{kT}{6\pi\eta a}. \quad (4.13)$$

For a sphere of radius $a = 10^{-4}$ cm in water at room temperature, $f_{sphere} = 1.9 \times 10^{-5}$ g/sec, and $D_{sphere} = 2.2 \times 10^{-9}$ cm^2/sec.

The net force acting on a liquid droplet is slightly smaller than that for a solid sphere of the same size, because the liquid can flow backward along the sides of the droplet (in the $+\theta$ direction, Fig. 4.4) and return along its axis. This reduces the shear in the external medium. For a sphere of viscosity η' in a medium of viscosity η,

$$f_{droplet} = 6\pi\eta a \frac{2\eta + 3\eta'}{3\eta + 3\eta'}; \quad (4.14)$$

see §337 of Lamb (1932). In the limit $\eta' \to \infty$, this reduces to Eq. 4.12. In the limit $\eta' \to 0$, we get the frictional drag coefficient for a bubble, e.g., of air:

$$f_{bubble} = 4\pi\eta a. \quad (4.15)$$

Note that a spherical cell behaves as a solid rather than as a liquid, because the plasma membrane is not free to flow through the middle of the cell. The same thing is true for a droplet or bubble in a medium containing surface-active agents that form a monolayer at the interface.

The frictional drag coefficients of a disk and an ellipsoid of revolution are compared in Fig. 4.5. When a particle diffuses, it continuously changes its orientation. The frictional drag coefficient, $\bar{f}$, that characterizes the average drift velocity (and average diffusion coefficient) of

Diffusion with Drift—57

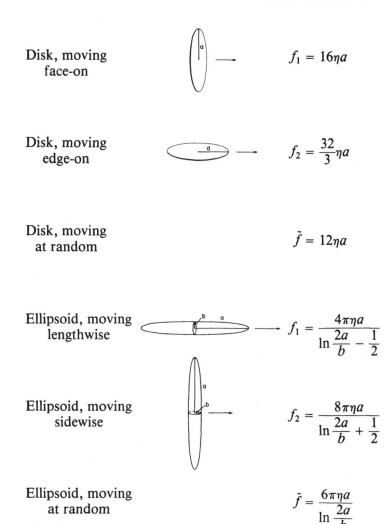

Disk, moving face-on		$f_1 = 16\eta a$
Disk, moving edge-on		$f_2 = \dfrac{32}{3}\eta a$
Disk, moving at random		$\bar{f} = 12\eta a$
Ellipsoid, moving lengthwise		$f_1 = \dfrac{4\pi\eta a}{\ln\dfrac{2a}{b} - \dfrac{1}{2}}$
Ellipsoid, moving sidewise		$f_2 = \dfrac{8\pi\eta a}{\ln\dfrac{2a}{b} + \dfrac{1}{2}}$
Ellipsoid, moving at random		$\bar{f} = \dfrac{6\pi\eta a}{\ln\dfrac{2a}{b}}$

Fig. 4.5. Comparison of the viscous drag coefficients of a solid circular disk of radius a and a prolate ellipsoid of revolution of semimajor axis a and semi-minor axes b. The expressions for the ellipsoid are valid in the limit $a^2 \gg b^2$. For general expressions for oblate and prolate ellipsoids, see p. 499 of Perrin (1934). The average coefficient $\bar{f}$ was obtained from f_1 and f_2 as described in the text; see also pp. 10–11 of Perrin (1936). The frictional drag coefficient for a solid sphere of radius a is $6\pi\eta a$, Eq. 4.12.

such a particle is given by $1/\bar{f} = (1/f_1 + 1/f_2 + 1/f_3)/3$, where f_1, f_2, and f_3 are the drag coefficients for motion along the principal axes. For the particles shown in Fig. 4.5, $f_2 = f_3$. Note that even though a particle is highly asymmetric, f_1 and f_2 differ by less than a factor of 2.

Stokes' law gives a good ballpark estimate for the viscous drag on globular things, even for particles as small as a molecule of the protein lysozyme. As a first approximation, do not worry about the shape of the particle, just think in terms of a sphere of roughly the same linear size. However, situations do arise in which asymmetries in viscous drag matter. One that we will encounter in Chapter 6 is flagellar propulsion. A great deal of effort has gone into figuring out the hydrodynamic properties of objects of complex shape; see, for example, García de la Torre and Bloomfield (1981).

Sedimentation rate

We are now in a position to write down the equations governing the sedimentation of particles in a gravitational or centrifugal field. A particle of mass m and volume V is suspended in a fluid of specific gravity ρ and viscosity η and subjected to a gravitational field of acceleration g, as shown in Fig. 4.6. The net downward force on the particle is

$$F_{down} = m'g, \qquad (4.16)$$

where m' is the effective mass of the particle, the mass of the particle less the mass of the fluid it displaces:

$$m' = m - V\rho. \qquad (4.17)$$

In a vacuum, the downward force on the particle would be mg. In a fluid, it is reduced by an amount $V\rho g$, the force due to buoyancy. If $V\rho$ is smaller than m (m' posi-

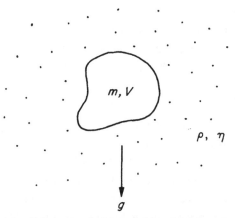

Fig. 4.6. A particle suspended in a fluid and subjected to a gravitational field. The particle has mass m and volume V. The fluid has specific gravity ρ, viscosity η. The downward acceleration is g.

tive), the particle sinks; if it is larger than m (m' negative), the particle floats; if it is equal to m (m' zero), the particle is neutrally buoyant. Equation 4.17 often is expressed as $m' = m(1 - \bar{v}\rho)$, where $\bar{v}$ is the specific volume of the particle, V/m.

From Eqs. 4.2 and 4.3 we have

$$v_d = \frac{F_{down}}{f} = \frac{m'gD}{kT}. \quad (4.18)$$

This is the Svedberg equation. It usually is written

$$\frac{v_d}{g} = S = \frac{m'D}{kT}, \quad (4.19)$$

where S is the sedimentation rate per unit acceleration. The units of S are in sec; 1 Svedberg = 10^{-13} sec. A 70 S particle, such as a ribosome from the bacterium *Escherichia coli*, sediments in a field of 1 cm/sec² at the rate 70×10^{-13} cm/sec. The acceleration due to gravity, g, is about 980 cm/sec², so in an ultracentrifuge at

100,000 × g, the 70 S particle sediments at the rate 70×10^{-5} cm/sec. This velocity is only 10^{-5} as large as the instantaneous root-mean-square velocity of such a particle, which is about 100 cm/sec. The centrifugal field adds a small but persistent bias to the much more riotous motion due to thermal energy.

Note that the sedimentation rate depends both on the effective mass, m', and on the diffusion coefficient, D (or the frictional drag coefficient, f). If two particles have the same effective mass, the one that is more compact sinks more rapidly. To cite an extreme example, a man wearing a parachute reaches a much higher terminal velocity when his chute fails to open than when it functions properly, even though his effective mass is the same in either case.

In a centrifuge, we deal not with the gravitational acceleration g, but with a centrifugal acceleration $r\omega^2$, where r is the distance from the axis of rotation and ω is the angular velocity of the rotor in radians/sec (2π times the rotation rate in revolutions per sec). The sedimentation rate increases with distance from the axis of rotation, because r increases, but for now we ignore this complication. In general, there are two ways of doing an experiment. Consider a solution containing two kinds of particles whose sedimentation rates differ by a factor of about two. Either we start with a centrifuge tube (or sector cell) filled with the mixture, as shown at the top of Fig. 4.7, or we layer a small sample at the top of a tube containing something else, usually a density gradient of sucrose, as shown at the bottom of Fig. 4.7. In the former case, the particles with the smaller sedimentation rate lag behind near the top of the tube. In the latter case, the two species separate out into different bands. The sucrose gradient is absolutely essential; without it, the bands would have larger specific gravities than the fluid beneath them, and they would sink in bulk by convective flow. At

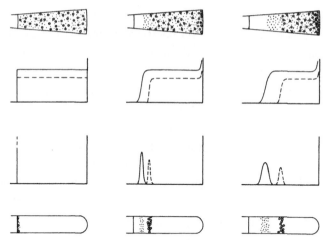

Fig. 4.7. Sedimentation rate experiments involving a mixture of two kinds of particles having sedimentation rates that differ by a factor of about two, shown at times 0 (left), t_0 (middle), and $2t_0$ (right). In the upper experiment, the mixture initially fills a sector cell in an analytical centrifuge. In the lower experiment, it is layered at the top of a sucrose gradient in a swinging bucket of a preparative centrifuge. The same centrifugal field is applied in either case; the particles sediment to the right. The upper and lower graphs show the concentrations of particles observed in the two experiments. The lower graph also depicts $\partial C/\partial r$, where C is the concentration in the upper experiment. In the analytical centrifuge, this function can be displayed directly with schlieren optics. Note in the upper experiment that the particles pile up at the bottom of the cell.

higher sucrose densities, the particles sediment more slowly, but we ignore this buoyant effect. Diffusion broadens the moving boundaries shown in the upper graph of Fig. 4.7, in accordance with Eq. 2.13 (see Fig. 2.5), and it broadens the moving bands shown in the lower graph, in accordance with Eq. 1.22 (see Fig. 1.3). The relative displacement of the boundaries (or bands) increases linearly with time, while the spreading increases only as the square-root of the time; therefore, the separation improves with the square-root of the time. One can always improve the separation by working at higher fields, because this increases the sedimentation rates but does not change the diffusion coefficients. From the spreading, one can compute D; from the sedimentation rate, given D, one can compute the effective mass, m'. If the specific volume of the particle is known, one can use Eq. 4.7 and compute the mass, m.

From the spreading shown in Fig. 4.7, it is evident that more rapidly sedimenting particles have smaller diffusion coefficients. This is generally true for globular particles made of a similar material. The effective mass of a sphere of radius a and specific gravity ρ_s is

$$m'_{sphere} = m - V\rho = V(\rho_s - \rho) = 4\pi a^3(\rho_s - \rho)/3.$$

(4.20)

The sedimentation rate of the sphere is

$$v_{d,sphere} = \frac{F_{down}}{f} = \frac{m'g}{6\pi \eta a} = 2a^2(\rho_s - \rho)g/9\eta,$$

(4.21)

a quantity that increases as a^2. As we already have seen in Eq. 4.13, the diffusion coefficient of the sphere is inversely proportional to a. Therefore, a sphere that sediments twice as fast has a diffusion coefficient that is smaller by a factor of $2^{-1/2}$. A sphere of radius $a = 10^{-4}$ cm and specific gravity $\rho_s = 1.2$ g/cm^3 sediments in water in a field $1 \times g$ at the rate $v_{d,sphere} = 4.4 \times 10^{-5}$ cm/sec.

Electrophoresis

If a particle carries an electric charge, then one can exert a force on it with an electric field. An ion carrying charge q (esu) in an electric field of intensity E (statvolts/cm) experiences a force in the direction of the field Eq (dynes). Unfortunately, q is not easy to define. Particles of biological interest contain a variety of ionizable groups whose charges depend strongly on pH. These charges are shielded by counter-ions attracted from the medium in which the particles are suspended. The effectiveness of the shielding depends on the ionic strength. So you do not hear much about particles that have specified electrophoretic drift rates per unit field (as you do, for example, about 30, 50, or 70 S ribosomes). Nevertheless, electrophoretic methods of separating and characterizing biological materials are extremely useful. In practice, they are remarkably simple.

As in the case of sedimentation, Fig. 4.7, there are essentially two ways of doing an experiment. Either one creates a sharp interface between a column of liquid containing a mixture of the particles to be studied and a column of liquid devoid of such particles and then passes an electrical current from one to the other, generating patterns analogous to those shown at the top of Fig. 4.7, or one layers a mixture of particles at the top of a medium designed to suppress convective stirring and passes an electrical current through that, generating patterns analogous to those shown at the bottom of Fig. 4.7. The physics is much the same: the relative displacement of the boundaries (or bands) increases linearly with time, while the spreading increases as the square-root of the time; so the separation improves as the square-root of the time. But in the case of electrophoresis, it is not always possible to improve the separation by increasing the field, because the electric current generates heat. The heat is generated at the same rate at all points across a transverse section of the electrophoretic column, but it is dissipated only at the edges, so the center of the column becomes relatively hotter. In most of the media used to suppress convection (see below), this increases the electrophoretic mobility, and the bands become curved.

As noted earlier, convective stirring is suppressed in the ultracentrifuge by the use of density gradients, e.g., of sucrose or CsCl. In an electrophoresis experiment, it is more convenient to use a gel, e.g., polyacrylamide or agarose. At the end of the experiment the bands can be precipitated into the gel and/or stained, e.g., with colored or fluorescent dyes, or the gel can be dried down and exposed to X-ray film to reveal components that are radioactive. Gels are not used in the ultracentrifuge, because they collapse in large centrifugal fields.

Gels not only suppress convective stirring, they act as molecular sieves. The rate of migration of a particle through the gel is strongly dependent on size. A particle that is small compared with the pores in the gel can diffuse through it, almost as if the gel were not there. A particle that is large compared with the pores in the gel simply is immobilized. Particles of intermediate size get through with varying degrees of difficulty. Particles that would move through a dilute aqueous medium at roughly the same rate move through the gel at rates that decrease exponentially with size; as a result, an estimate of size (or mass) can be made from a measurement of the logarithm of the displacement. Pieces of DNA and RNA are routinely sorted in this way, as are proteins dissolved in ionic detergents, such as sodium dodecyl sulfate. It is easy to distinguish gels of this kind, because the faster-moving bands always are broader; the molecules that drift more rapidly are smaller and have larger diffusion coefficients.

Chapter 5

Diffusion at Equilibrium

What happens if the sedimentation-rate experiment shown at the top of Fig. 4.7 is continued for a very long time? Eventually, the particles will pile up near the bottom of the cell and form a distribution that does not change. The particles still will diffuse up and down, but the average number at any given point will remain constant. It is often argued that equilibrium will be established when the upward flux of particles due to diffusion balances the downward flux due to sedimentation, the equilibrium concentration, C, being the steady-state solution of Eq. 4.5, with v_d given by Eq. 4.18. This is true but misleading. One ought to be able to figure out what the equilibrium distribution is without having to appeal to fluxes, quantities that involve the movement of particles with time. The equilibrium distribution is, by definition, time invariant. In any event, if we were to work out the equilibrium distribution from Eq. 4.5, the argument would be circular, because we derived Eq. 4.18 from Eq. 4.3 following an appeal to Eq. 1.1, which refers to particles at equilibrium. Evidently, the answer we seek and Eq. 1.1 are intimately related.

For the sake of simplicity, consider the concentration of particles at height x above a horizontal barrier in a uniform field, g, as shown in Fig. 5.1. At equilibrium, there will be fewer particles at large values of x than at small values of x, just as there are fewer molecules of air at high altitudes than at the surface of the earth. The work required to lift a particle a distance x is $m'gx$, the downward force of Eq. 4.16 times the distance x. Thus,

66—Diffusion at Equilibrium

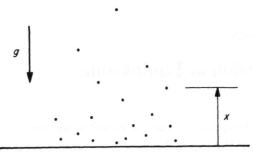

Fig. 5.1. Equilibrium distribution of particles of effective mass m' above a horizontal barrier in a field of uniform acceleration g. The height above the barrier is x. See Eq. 5.4.

particles at large values of x have larger gravitational energies. Given that the particles are able to move around and interact with the molecules of the fluid in which they are suspended, what is the probability that a particle will be found with gravitational energy $m'gx$? Or to state the problem more generally, given that the particles can exist in states 1, 2, ..., i, ..., having different energies E_1, E_2, ..., E_i, ..., what is the probability that a given particle will be found in state i? The answer is

$$P_i = \frac{1}{Q} e^{-E_i/kT}, \tag{5.1}$$

where

$$Q = \sum_i e^{-E_i/kT} \tag{5.2}$$

is a normalization constant called the partition function, the summation extending over all possible states i. If the energy levels are continuous rather than discrete, the sum becomes an integral. Eq. 5.1 is the Boltzmann equation, the basic equation of classical equilibrium statistical mechanics. If there are N particles in all, $N_i = NP_i$ particles will be found in state i. If we are interested in the ratio of the probability of finding a particle in state i to the probability of finding a particle in state j, or in the ratio of the number of particles in state i to the number of particles in state j, then Eq. 5.1 implies that

$$\frac{N_i}{N_j} = \frac{P_i}{P_j} = e^{-(E_i-E_j)/kT}, \qquad (5.3)$$

and we need not concern ourselves with Q. For the particles of Fig. 5.1, Eq. 5.3 takes the form

$$\frac{C(x)}{C(0)} = \frac{N(x)}{N(0)} = e^{-m'gx/kT}. \qquad (5.4)$$

The concentration of particles decreases exponentially with height, by a factor e in a distance $kT/m'g$. We will refer to this distance as the scale height. For molecules of air over the earth, the scale height is about 8 km.

A particle in a Boltzmann distribution still diffuses. It moves up and down, spending most of its time near $x = 0$, but sometimes wandering several scale heights away. Equation 5.4 tells us the probability that the particle will be found at any given height x. If there are a large number of similar particles, then Eq. 5.4 tells us the average number that will be found at any given height x. The mean height is just the scale height, x_s. Figure 5.2 shows the vertical displacement of a particle diffusing in a Boltzmann distribution. The plot extends over a period of time equal to 10 times the period required for the particle to diffuse a root-mean-square distance equal to its scale height.

68—Diffusion at Equilibrium

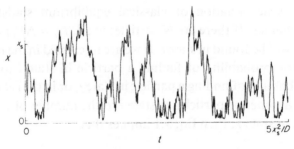

Fig. 5.2. The height of a particle in a Boltzmann distribution of scale height x_s shown for a period of time $10x_s^2/2D$. The motion was simulated on a computer by dividing a segment of the x axis into 200 equal units, assigning the particle a scale height of 40 units, and allowing it to step up or down one step at a time 16,000 times. Sedimentation was accounted for by making the probability of stepping up smaller than the probability of stepping down by a factor $e^{-1/40}$; at $x = 0$ the particle only was allowed to diffuse up. This ensures that the particle will be found at height x with probability e^{-x/x_s}. For the period of time shown, the particle spent most of its time at a height smaller than the scale height, on the average at about $0.6x_s$.

Derivation of the Boltzmann equation

Consider an ensemble of N particles with total energy E. Assume that N_i particles are in a state with energy E_i, N_j particles are in a state with energy E_j, etc., subject to the constraints $\sum_i N_i = N$ and $\sum_i N_i E_i = E$. If N is large, there are many different ways of parceling out this energy, i.e., of assigning particles to states. In classical statistics, the particles are distinguishable; for a given set of occupation numbers $[N_i] = N_1, N_2, \ldots, N_i, \ldots$, there are

$$W = \frac{N!}{N_1! N_2! \cdots N_i! \cdots} \tag{5.5}$$

different ways of assigning particles to states. To see this, note that if all the particles were in different states, a different assignment would arise every time we exchanged any two particles, i.e., for every one of the $N!$ possible permutations in the order of the particles. However, if there are N_1 particles in state 1, N_2 particles in state 2, etc., this number must be reduced by a factor $N_1!N_2!\ldots$, because the exchange of two particles in the same state does not change the assignment, and $N_1!N_2!\ldots$ is the total number of assignments that differ by such exchanges.

We now take a leap of faith and postulate that every possible way of parceling out the energy is, a priori, equally probable. If a given set of occupation numbers $[N_i]$ turns out to give a number of assignments that is enormous compared with that possible with any other set of occupation numbers, then the set of occupation numbers $[N_i]$ will be the one observed in practice. Although we will not prove it, this turns out to be the case. The problem is reduced to one of finding $[N_i]$ such that W is a maximum, subject to the constraints $\sum_i N_i = N$ and $\sum_i N_i E_i = E$, where N and E are constants. The result is

$$\frac{N_i}{N} = \frac{e^{\beta E_i}}{\sum_i e^{\beta E_i}}, \quad (5.6)$$

where β is a constant.

One can show that β is solely a function of temperature by considering two ensembles of particles N and N' separated by a barrier that blocks the passage of matter but allows an exchange of energy. The number of different ways, $W + W'$, of assigning particles to states, subject to the constraint that the total energy $E + E'$ is constant, is maximum when both sets of occupation numbers satisfy Eq. 5.6, with the *same* value of β. It is a matter of common experience (and thermodynamic postulate) that different systems in thermal contact come to equilibrium at the same temperature; therefore, β is a measure of temperature. By using Eq. 5.6 to compute the expectation value of a suitable macroscopic parameter, such as the pressure of an ideal gas, it is possible to show that $\beta = -1/kT$. Thus Eq. 5.6 leads to Eq. 5.1.

This is a remarkable result. It follows from the assumption that every possible microscopic description of a system of N particles of energy E is equally probable. If N is large, essentially all these descriptions involve the set of occupation numbers specified by the Boltzmann equation; the most probable set is overwhelmingly most probable. If a particle has different degrees of freedom—i.e., if it can translate along the x, y and z axes, rotate about these axes, vibrate internally, etc.—then the Boltzmann equation applies to each of these degrees of freedom separately. The total available energy is distributed among the different degrees of freedom in such a way that each satisfies the Boltzmann equation with the same value of the temperature T.

In quantum statistics, the particles are indistinguishable, and assignments such as those made in Eq. 5.5 that differ by the exchange of two particles are not counted separately. The occupation number of a state of a particle of half-integral spin, such as an electron, can be either 0 or 1: this leads to Fermi-Dirac statistics. The occupation number of a state of a particle of integral spin, such as a photon, can be anything from 0 to N: this leads to Bose-Einstein statistics. In other respects, the derivations are similar to that of Boltzmann statistics.

The importance of kT

At thermal equilibrium, the probability of finding a particle in a state with an energy large compared with kT is quite small. For example, the probability of finding a particle in a state of energy $10kT$ or more is only 4.5×10^{-5}. However, the probability of finding a particle in a state with an energy of the same order of magnitude as kT is quite large. The probability of finding a particle in a state of energy kT or more is 0.37. Thus, kT can be thought of as the energy of thermal fluctuation. When a particle interacts with its surroundings at temperature T, it frequently gains and loses amounts of energy of this order of magnitude. This is evident, for example, in the changes of height of a particle that has sedimented to equilibrium, as illustrated in Figs. 5.1 and 5.2.

Mean-square velocity

A particle of mass m moving along the x axis at velocity v_x has a kinetic energy $mv_x^2/2$. Equation 5.1 implies that

$$\langle v_x^2 \rangle = \int_0^\infty v_x^2 P(v_x) dv_x$$

$$= \frac{\int_0^\infty v_x^2 e^{-mv_x^2/2kT} dv_x}{\int_0^\infty e^{-mv_x^2/2kT} dv_x}. \quad (5.7)$$

On consulting a table of definite integrals, we find

$$\langle v_x^2 \rangle = \frac{(\pi/16)^{1/2} (2kT/m)^{3/2}}{(\pi/4)^{1/2} (2kT/m)^{1/2}} = \frac{kT}{m}, \quad (5.8)$$

the value asserted in Eq. 1.1.

Einstein-Smoluchowski relation

We used Eq. 1.1 in deriving the Einstein-Smoluchowski relation, Eq. 4.3. A more general proof is now possible via Eq. 4.4, which, given Eq. 4.2, takes the form

$$J_x = -D\frac{\partial C}{\partial x} + \frac{F_x}{f} C. \quad (5.9)$$

72—Diffusion at Equilibrium

The forces of interest, F_x, are conservative: the work that they do in moving a particle from position 1 to position 2 is independent of the path taken by the particle; it only depends on the end points. Such a force can be expressed as a gradient of a potential energy, $F_x = -\partial E/\partial x$, so that

$$J_x = -D\frac{\partial C}{\partial x} - \frac{C}{f}\frac{\partial E}{\partial x}. \tag{5.10}$$

At equilibrium, $J_x = 0$, and Eq. 5.10 implies that

$$\frac{dC}{C} = -\frac{1}{Df} dE. \tag{5.11}$$

Integrating this equation, we find

$$\frac{C(x)}{C(0)} = e^{-[E(x)-E(0)]/Df}, \tag{5.12}$$

which on comparison with Eq. 5.3 requires that $D = kT/f$, the desired result.

Equilibrium sedimentation

Given the Boltzmann equation, we know at once the equilibrium distribution of particles in a gravitational field, Eq. 5.4. Things are slightly more complicated in the centrifuge, because the downward acceleration, $r\omega^2$, varies with r. The work required to lift the particle from radius r_0, the bottom of the cell, to radius r, is

$$E(r) - E(r_0) = -\int_{r_0}^{r} m'r\omega^2 dr$$

$$= m'\omega^2(r_0^2 - r^2)/2, \tag{5.13}$$

so that

$$\frac{C(r)}{C(r_0)} = e^{-m'\omega^2(r_0^2-r^2)/2kT}. \tag{5.14}$$

Density-gradient sedimentation

If particles are suspended in a dense salt, e.g., in a solution of CsCl or $CsSO_4$, and the experiment is run long enough at high enough fields, then the salt will sediment to equilibrium, and the density of the fluid (water + salt) will be larger near the bottom of the cell than near the top. In practice, the difference is not large, because the scale height of the distribution of salt is larger than the height of the column of liquid. Under these conditions, the particles will form a band centered at their own buoyant density, m/V. Particles above this point will sediment downward; particles below this point will sediment upward. How broad is the band? What is its shape? We return to the integral in Eq. 5.13 and note that m' is now a function of r, since the density of the fluid, ρ, is a function of r. Suppose the band is narrow and centered at $r = r_b$. The density of the fluid near r_b can be expressed in a Taylor expansion as

$$\rho(r) = \rho(r_b) + \frac{\partial \rho}{\partial r}\bigg|_{r=r_b} (r - r_b)$$

+ terms of order $(r - r_b)^2$ and higher. (5.15)

By substituting this approximate value of $\rho(r)$ into Eq. 5.13, noting that $1 - \bar{v}\rho(r_0) = 0$, evaluating the integral, and using it in the Boltzmann equation, we find

$$\frac{C(r)}{C(r_b)} = e^{-mr_b\omega^2\bar{v}\rho'(r_b)(r-r_b)^2/2kT}, \quad (5.16)$$

where $\rho'(r_b) = \partial\rho/\partial r|_{r=r_b}$. This is a Gaussian distribution (compare Eq. 1.22) with standard deviation

$$\sigma_r = [kT/mr_b\omega^2\bar{v}\rho'(r_b)]^{1/2}. \quad (5.17)$$

Thus, the band is narrow for particles of large mass (large m), in high fields (large $r_b\omega^2$), and in steep gradients (large

$\rho'(r_b)$). If the material under study is inhomogeneous, the band will be broader, with particles of lower mass predominating at larger values of $(r - r_b)$. The method is particularly useful for isolating large molecules of a given kind that come in different sizes or tend to fragment, such as DNA, since particles with the same buoyant density band in the same region, regardless of their size.

Isoelectric focusing

An analogous situation arises in electrophoresis when the experiment is run in a pH gradient. At equilibrium, a protein will form a band centered at the pH at which it is electrically neutral, i.e., at its isoelectric point. A particle at a more acid pH is positively charged and moves toward the cathode; a particle at a more basic pH is negatively charged and moves toward the anode. Thus, the pH gradient must be acidic near the anode and basic near the cathode. The band is narrow for particles with large numbers of titratable groups, in high fields, and in steep gradients. The gradient is established by the migration of an array of polyelectrolytes that are electrically neutral at different values of pH. These ampholytes play a role in isoelectric focusing analogous to that of the heavy salts in density-gradient sedimentation.

Chapter 6

Movement of Self-propelled Objects

The drift imposed on the diffusion of particles discussed in Chapter 4 was due to forces externally applied: to gravitational, centrifugal, or electrical fields. There are many microscopic objects in this world that are self-propelled. Among the most remarkable of these are flagellated bacteria: spherical, rod-shaped, or helical cells as small as 10^{-4} cm in diameter that swim steadily at speeds of order 2×10^{-3} cm/sec. While it may be difficult to imagine oneself as an inert particle falling in a centrifugal field, it is less so to contemplate the plight of a tiny living thing. What is it like to be so small?

Life at low Reynolds number

In the discussion leading to Stokes' law, we noted that when small things move through fluids slowly, viscous forces are important but inertial forces are not. A hydrodynamicist would say that we are dealing with fluids at low Reynolds number. The Reynolds number is a dimensionless parameter in the equations of motion of a fluid that indicates the relative size of terms that describe inertial forces (forces required to accelerate masses) and viscous forces (forces due to viscous shear). The Reynolds number is

$$R \simeq \frac{vL\rho}{\eta}, \qquad (6.1)$$

where v is the velocity of the fluid (the velocity of a particle moving through the fluid), L is the linear scale of the motion (the size of the particle), ρ is the specific gravity of the fluid, and η is its viscosity. For a fish, $v \simeq 10^2$ cm/sec, $L \simeq 10$ cm, $\rho \simeq 1$ g/cm^3, and $\eta \simeq 10^{-2}$ g/cm sec, so that $R \simeq 10^5$. For a bacterium, $v \simeq 10^{-3}$ cm/sec, $L \simeq 10^{-4}$ cm, $\rho \simeq 1$ g/cm^3, and $\eta \simeq 10^{-2}$ g/cm sec, so that $R \simeq 10^{-5}$. The Reynolds number of the fish is very large, that of the bacterium is very small. The fish propels itself by accelerating water, the bacterium by using viscous shear. The fish knows a great deal about inertia, the bacterium knows nothing. In short, the two live in very different hydrodynamic worlds.

To make this point clear, it is instructive to compute the distance that a bacterium can coast when it stops swimming. To do so, we approximate the cell as a sphere of radius a and specific gravity ρ_s, as shown in Fig. 6.1. According to Newton's second law (force = mass × acceleration), the sphere decelerates according to the equation

$$m\left(-\frac{dv}{dt}\right) = 6\pi\eta a v,$$

where m is the mass of the sphere, $4\pi a^3 \rho_s/3$. Rearranging,

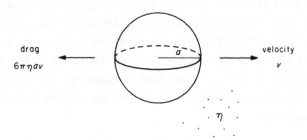

Fig. 6.1. A sphere of radius a and specific gravity ρ_s moving to the right at velocity v through a fluid of viscosity η.

we have

$$\frac{dv}{v} = -\frac{6\pi\eta a}{m} dt.$$

This differential equation has the solution

$$v(t) = v(0)e^{-t/\tau},$$

where $v(0)$ is the initial velocity of the sphere and

$$\tau = \frac{m}{6\pi\eta a} = \frac{2a^2\rho_s}{9\eta}.$$

For $a = 10^{-4}$ cm, $\rho_s = 1$ g/cm^3, and $\eta = 10^{-2}$ g/cm sec, $\tau \simeq 2 \times 10^{-7}$ sec; the bacterium comes to a stop in about 1 μsec! In doing so, it coasts a distance

$$d = \int_0^\infty v(t)dt = v(0)\tau.$$

A cell moving at an initial velocity of 2×10^{-3} cm/sec coasts 4×10^{-10} cm = 0.04 Å, a distance small compared with the diameter of a hydrogen atom!! Note that the bacterium is still subject to Brownian movement, so it does not actually stop. The drift goes to zero, not the diffusion.

The power required to push such a cell is actually quite small. It is equal to the viscous drag times the velocity: for a sphere of radius a, $6\pi\eta av^2$. For the cell just considered, the power is about 8×10^{-18} watts (8×10^{-11} ergs/sec). For a cell burning glucose, this corresponds to the consumption of oxygen at the rate of about 25 molecules/sec. But when a small bacterium grows in a rich medium, it utilizes oxygen at a rate about 10^5 times as large as this. Things are not so favorable when we realize that the efficiency of the propulsion system is at best a few percent, but for most bacteria energy supplies are ample.

For deeper insights into life at low Reynolds number, see the article with this title by Purcell (1977).

Flagellar propulsion

The net force on a cell swimming at a constant velocity is zero. If the net force were not zero, the cell would accelerate or decelerate. This is true regardless of the nature of the forces, i.e., regardless of the size of the organism. Flagellated bacteria swim by rotating one or more thin helical filaments that extend out into the external medium, as shown in Fig. 6.2. Torque generated by rotation of the filaments is balanced by viscous drag due to counter-rotation of the body of the cell, and thrust generated by rotation of the filaments is balanced by viscous drag due to translation of the body of the cell. It is easy to see that torque is required to rotate a helical filament in a viscous medium, but it is not obvious why this rotation should generate thrust. The answer lies in the fact that the viscous drag on a thin rod or filament is larger when the filament moves at a given velocity sideways than when it

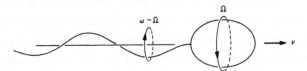

Fig. 6.2. A bacterium swimming to the right at velocity v. A rotary motor at the base of the flagellum turns the helical filament relative to the body of the cell at angular velocity ω. When the cell is viewed from behind, the body turns clockwise at angular velocity Ω, and the filament turns counterclockwise at angular velocity $\omega - \Omega$; ω is larger than Ω. The filament is left-handed; when the helix is viewed end-on, a particle moving along it away from the observer turns counterclockwise.

moves end-on. As we saw for the thin prolate ellipsoid, Fig. 4.5, the difference approaches a factor of 2. To get the gist of the argument, consider the viscous drag on two segments of the flagellar helix shown in Fig. 6.3. The segments are half a turn out of phase. As discussed in the legend of the figure, the components of the drag normal to the axis of the helix, F_Ω and F_Ω', contribute to the torque, while the components of the drag parallel to the axis of the helix, F_v and F_v', contribute to the thrust. If there is an integral number of turns, the net force normal to the axis of the helix is zero. If there is not, the cell moves with a conical wobble.

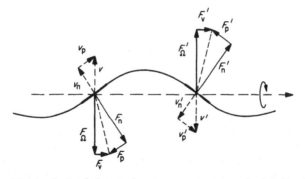

Fig. 6.3. Analysis of viscous drag on two segments of a flagellar filament moving slowly to the right and turning rapidly counterclockwise. The velocity of each segment, v, is decomposed into velocities normal and parallel to the segment, v_n and v_p, respectively. The segment shown on the left is moving upward in front of the plane of the paper; the one shown on the right (denoted by primes) is moving downward behind the plane of the paper. The frictional drags normal and parallel to each segment, F_n and F_p, act in directions opposite to v_n and v_p, respectively. Note that their magnitudes are in the ratios $F_n/F_p \simeq 2v_n/v_p$. F_n and F_p are decomposed into components normal and parallel to the helical axis, F_Ω and F_v, respectively. F_Ω and F_Ω' act in opposite directions and form a couple that contributes to the torque. F_v and F_v' act in the same direction and contribute to the thrust.

80—Movement of Self-propelled Objects

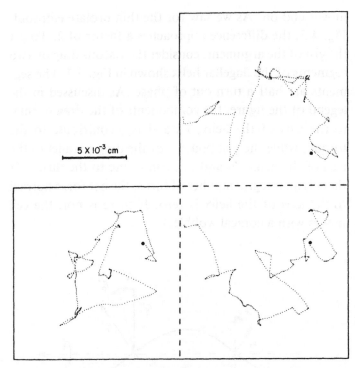

Fig. 6.4. A digital plot of the displacement of a wild-type bacterium, *E. coli* strain AW405, executing a random walk in a homogeneous, isotropic medium. The plots are planar projections of a three-dimensional path. If the left and upper panels are folded out of the page along the dashed lines, the projections appear in proper orientation on three adjacent faces of a cube. Observation began at the end of the path denoted by the large dot and continued for about 30 sec. There are 12.6 dots/sec; the bar is 5×10^{-3} cm long. The cell swam at a speed of about 2×10^{-3} cm/sec. Run intervals were relatively long, about 1 sec on the average, and tumble intervals were relatively short, about 0.1 sec on the average. There were 26 runs and tumbles. The tracking was done at 32°C in a medium of viscosity 0.027 g/cm sec. The data are from Berg and Brown (1972). A stereo pair of this track is shown in Fig. 7 of Berg (1978).

Motility of *Escherichia coli*

Flagellated bacteria swim in a manner that depends on the size and shape of the body of the cell and the number and distribution of the flagella. The common intestinal organism *Escherichia coli* looks like a small cocktail sausage. It has a rod-shaped body about 10^{-4} cm in diameter by 2×10^{-4} cm long. Approximately six flagellar filaments emerge at random points on the sides of the body and extend about three body lengths into the surrounding medium. When these flagella turn counterclockwise (as defined in Fig. 6.2), they form a synchronous bundle that pushes the body steadily forward; the cell is said to "run." When they turn clockwise, the bundle comes apart and the flagella turn independently, moving the cell body this way and that in a highly erratic manner; the cell is said to "tumble." These modes alternate, and the cell executes a three-dimensional random walk, as shown in Fig. 6.4. For a description of the different flagellar motions that occur during runs and tumbles, see Macnab and Ornston (1977).

Rotational diffusion

One striking thing about the runs shown in Fig. 6.4 is that they are not straight. The cell meanders; it changes direction during a run, on the average, about 27°. Without external clues, the cell has no absolute sense of direction. It is subject to rotational diffusion. In Chapter 1 we noted that microscopic particles have an average kinetic energy associated with translation along any axis of $kT/2$. They also have an average kinetic energy associated with rotation about any axis of $kT/2$. Rotational diffusion can be analyzed in much the same fashion as translational diffusion. Instead of stepping a distance $\pm\delta$ along the x axis every τ seconds, as shown in Fig. 1.2, we step an angle $\pm\phi$ about the x axis, as shown in Fig. 6.5. The mean-square angular deviation in time t is

$$\langle\theta^2\rangle = 2D_r t, \tag{6.2}$$

where $D_r = \phi^2/2\tau$ is the rotational diffusion coefficient (in units of radians^2/sec). Equation 6.2 is analogous to Eq. 1.10. The angular drift velocity, $\Omega_d = (d\theta/dt)_{drift}$, resulting from an externally applied torque about the θ axis, N_θ, is

$$\Omega_d = \frac{N_\theta}{f_r}, \tag{6.3}$$

where f_r is the rotational frictional drag coefficient. Equation 6.3 is analogous to Eq. 4.2. The analogue of

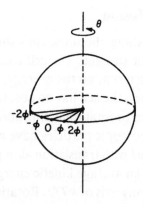

Fig. 6.5. Rotational random walk of a sphere about one axis in steps $\pm\phi$. See Eq. 6.2.

Eq. 4.3 is

$$D_r = \frac{kT}{f_r}. \qquad (6.4)$$

For a sphere of radius a,

$$f_{r,sphere} = 8\pi\eta a^3, \qquad (6.5)$$

so that

$$D_{r,sphere} = \frac{kT}{8\pi\eta a^3}. \qquad (6.6)$$

Equations 6.5 and 6.6 are analogous to Eqs. 4.12 and 4.13.

Rotational diffusion affects the precision with which a microorganism can swim in a straight line. A cell swimming along the x axis can deviate either by diffusing about the y axis or by diffusing about the z axis, as shown in Fig. 6.6; so it goes off course in a two-dimensional rotational

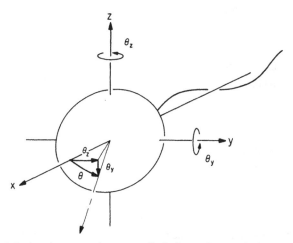

Fig. 6.6. A microorganism propelled along the x axis by a helical flagellum can wander off course by rotating about either the y or the z axis. The total angular deviation, θ, has components, θ_y and θ_z: $\theta^2 = \theta_y^2 + \theta_z^2$. Rotation about the x axis does not contribute to this deviation.

random walk, with $\langle \theta^2 \rangle = 4D_r t$. If we approximate the bacterium as a sphere of radius $a = 10^{-4}$ cm and use Eq. 6.6 with $T = 305°$ K and $\eta = 0.027$ g/cm sec (the conditions of the tracking experiment), we find $D_r \simeq 0.062$ radians^2/sec. This implies a root-mean-square angular deviation in radians of about $0.5 t^{1/2}$, where t is in seconds; so the cell wanders off course about 30° in 1 sec, as observed.

Rotational frictional drag coefficients of a disk and an ellipsoid of revolution are compared in Fig. 6.7. The rota-

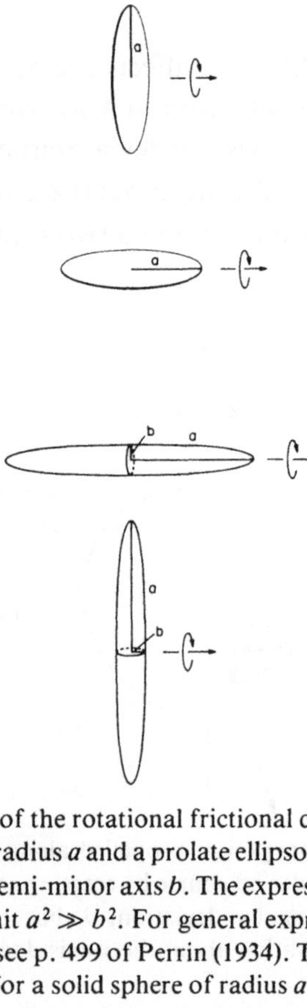

Disk, about an axis normal to the face through the center		$f_{r_1} = \dfrac{32}{3}\eta a^3$
Disk, about an axis parallel to the face through the center		$f_{r_2} = \dfrac{32}{3}\eta a^3$
Ellipsoid, about the major axis		$f_{r_1} = \dfrac{16}{3}\pi\eta a b^2$
Ellipsoid, about a minor axis		$f_{r_2} = \dfrac{8\pi\eta a^3/3}{\ln\dfrac{2a}{b} - \dfrac{1}{2}}$

Fig. 6.7. Comparison of the rotational frictional drag coefficients of a solid circular disk of radius a and a prolate ellipsoid of revolution of semi-major axis a and semi-minor axis b. The expressions for the ellipsoid are valid in the limit $a^2 \gg b^2$. For general expressions for oblate and prolate ellipsoids, see p. 499 of Perrin (1934). The rotational frictional drag coefficient for a solid sphere of radius a is $8\pi\eta a^3$, Eq. 6.5.

tional frictional drag coefficients of the disk are the same about the three principal axes, but those of the thin prolate ellipsoid differ: the coefficient for rotation about one of the minor axes is much larger than that for rotation about the major axis. This is not surprising, given the fact that the velocity of an element at distance r from the axis of rotation of an object spinning at angular velocity Ω is $r\Omega$. It takes a much larger torque to spin a long thin object at a given rate end-over-end than about its long axis. If you try such experiments, remember to work at low Reynolds number!

Random changes in direction

Tumbles are much shorter than runs, but they generate much larger changes in direction. For the data shown in Fig. 6.4, the average change during a tumble is 103°. Thus, the flagella must actively reorient the cell. If directions were chosen entirely at random, one would expect the mean angular deviation to be 90°. To see this, consider a cell swimming along the z axis at the center of a unit sphere, as shown in Fig. 6.8. Choosing a new direction at random is equivalent to dividing the surface of the sphere into small patches and allowing the cell to pick one at random. The probability that the angular deviation will lie between θ and $\theta + d\theta$ is the area of the sector between θ and $\theta + d\theta$, namely $2\pi \sin \theta \, d\theta$, divided by the area of the sphere, 4π; therefore

$$P(\theta) \, d\theta = \frac{1}{2} \sin \theta \, d\theta. \qquad (6.7)$$

By evaluating the appropriate integrals between 0 and π, one can show that $\langle \theta \rangle = \pi/2$ and $\langle \theta^2 \rangle = (\pi^2/2) - 2$. The

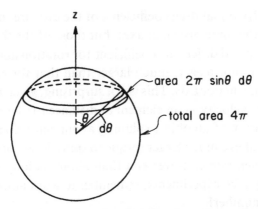

Fig. 6.8. A cell (not shown) swimming along the z axis of a unit sphere. If the cell chooses a new direction at random, the probability that it will change course by an angle between θ and $\theta + d\theta$ is $\frac{1}{2} \sin \theta \, d\theta$; see Eq. 6.7.

mean is 1.57 radians, or 90°; the standard deviation is 0.68 radians, or 39°.

Data from a large number of cells indicate that *E. coli*, when swimming in a medium of low viscosity, prefers to choose an angle somewhat less than 90° (68°, on the average); the experimental probability distribution is biased in the forward direction.

Poisson statistics

Another striking thing about the motion of *E. coli* is that the cells switch back and forth between run and tumble modes at random. The distribution of run (or tumble) intervals is exponential, and the length of a given interval does not depend on the lengths of the intervals that precede it. Evidently, the probability per unit time that a run (or tumble) will end is constant.

The Poisson interval distribution: Consider a process in which events of a single kind occur with a constant probability per unit time λ. If we look at the system at time $t = 0$, what is the probability that an event will be observed between time t and $t + dt$, where dt is infinitesimal? To find out, we divide the time interval t into a large number of increments, n, of length t/n, as shown in Fig. 6.9. The probability that an event occurs in the first increment is $\lambda t/n$. If n is large, this probability is small, and we need not worry about two or more events occurring in the same increment. The probability that an event does not occur in the first increment is $1 - \lambda t/n$. The probability that an event does not occur in n such successive increments, i.e., between times 0 and t, is $(1 - \lambda t/n)^n$. The probability that an event occurs in the next time increment dt is λdt. Therefore, the probability that the first event occurs between time t and $t + dt$ is

$$P(t;\lambda)dt = (1 - \lambda t/n)^n \lambda dt.$$

In the limit $n \gg 1$, the expansion of $(1 - \lambda t/n)^n$ as a binomial series gives the power series for $e^{-\lambda t}$; therefore

$$P(t;\lambda)dt = \lambda e^{-\lambda t}dt. \tag{6.8}$$

This is the Poisson interval distribution. As shown in Fig. 6.10, the shorter intervals are the more probable.

The expectation value of t^n is

$$\langle t^n \rangle = \int_0^\infty \lambda t^n e^{-\lambda t} dt = \frac{n!}{\lambda^n}. \tag{6.9}$$

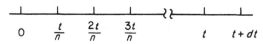

Fig. 6.9. A line extending from time 0 to time $t + dt$, divided into a large number of increments of length t/n.

88—Movement of Self-propelled Objects

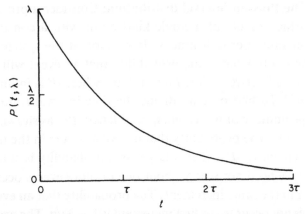

Fig. 6.10. The Poisson interval distribution, $P(t;\lambda)$, plotted in units of time $\tau = 1/\lambda$. See Eq. 6.8.

The mean interval $\langle t \rangle$ is $1/\lambda$. The mean-square interval $\langle t^2 \rangle$ is $2/\lambda^2 = 2\langle t \rangle^2$. Therefore, the standard deviation is $(\langle t^2 \rangle - \langle t \rangle^2)^{1/2} = \langle t \rangle$. The standard deviation is equal to the mean. If we measure a number of run intervals and compute the experimental mean and standard deviation, we do find that the standard deviation approximates the mean. This does not imply that we have made a poor measurement, as would be the case, say, had we been running replicate chemical assays; it is an inherent property of the Poisson process. However, if we were to repeat these measurements several times and plot the distribution of the experimental means, that distribution would be Gaussian, with a standard deviation of order $\langle t \rangle / n^{1/2}$, where n is the number of run intervals in each set of measurements.

The Poisson distribution: Instead of recording the intervals between events, suppose we count the number of events that occur in a fixed interval of time, t'. If we repeat this measurement a large number of times, how are these counts distributed? The probability that an event does not occur in the interval t' is 1 minus the probability that an event does occur, or

$$1 - \int_0^{t'} \lambda e^{-\lambda t} dt = e^{-\lambda t'}.$$

The probability that one event occurs in the interval t' is equal to the probability that an event occurs between t_0 and $t_0 + dt_0$ times the probability that an event does not occur in the remaining part of the interval, $t' - t_0$, integrated over all possible values of t_0, i.e.,

$$\int_0^{t'} \lambda e^{-\lambda t_0} e^{-\lambda(t'-t_0)} dt_0 = \lambda t' e^{-\lambda t'}.$$

The probability that two events occur in the interval t' is equal to the probability that an event occurs between t_0 and $t_0 + dt_0$ times the probability that one event occurs in the remaining part of the interval, $t' - t_0$, integrated over all possible values of t_0, i.e.,

$$\int_0^{t'} \lambda e^{-\lambda t_0} \lambda (t' - t_0) e^{-\lambda(t'-t_0)} dt_0 = \frac{(\lambda t')^2}{2} e^{-\lambda t'}.$$

Proceeding in this fashion, we find that the probability that k events occur in the interval of time t' is

$$P(k;\mu) = \frac{\mu^k}{k!} e^{-\mu}, \qquad (6.10)$$

where $\mu = \lambda t'$, and 0! is understood to be 1. This is the Poisson distribution discussed in Appendix A; see Eq. A.32. As indicated there, the mean of this distribution is μ, and its standard deviation is $\mu^{1/2}$.

Other processes that follow Poisson statistics: Poisson statistics are encountered whenever the probability that an event occurs in a given small increment of time or space is proportional to the size of the increment. The probability that an event occurs can be made arbitrarily small if the increments are made arbitrarily small. Thus, in the example considered in Fig. 6.9, the time interval t was divided into a large number of increments of length t/n, and the probability of the occurrence of an event in a given increment, $\lambda t/n$, approached 0 as $n \to \infty$, satisfying the criterion for passage from the binomial to the Poisson distribution discussed in Appendix A.

An example involving distributions in space rather than time is encountered in plating bacteria. We suspend N bacteria in a broth, volume V_0, and work with the suspension before the bacteria have had time to settle out. The probability per unit volume of finding a bacterium is $\lambda = N/V_0$. This number is the same for any aliquot of the suspension that we choose to examine. If we were to run the suspension through a device that flags the passage of each bacterium (e.g., a Coulter counter) and record the volume, V, between successive counts, these volumes would be distributed in accord with Eq. 6.8, with $t = V$. If we were to plate equal volumes, V', of the suspension, incubate the plates, and count the colonies, the number of colonies per plate would be distributed in accord with Eq. 6.10, with $\mu = \lambda V'$. In particular, the probability of finding no colonies on a plate would be $e^{-\mu}$, and the probability of finding k colonies on a plate would be $(\mu^k/k!)e^{-\mu}$. The mean number would be μ, with standard deviation $\mu^{1/2}$.

Another such process encountered in the laboratory is radioactive decay. If μ counts are collected over an arbitrary period of time and the count is repeated, the chances are pretty good that the second count will differ from the first by as much as $\mu^{1/2}$. If the mean is μ, the standard deviation is $\mu^{1/2}$. Usually, one is confronted with background counts due to cosmic rays, natural radioactivity, or contamination. One counts the sample in the presence of the background, obtaining a combined count μ_{ab}, and then the background alone, obtaining a count μ_b. One would like to know μ_a, the count that would be obtained for the sample alone. It is possible to show that the sum of two Poisson processes also is a Poisson process. Thus, the best estimate of the standard deviation of the combined count is $\mu_{ab}^{1/2}$. The best estimate of the standard deviation of the background count is $\mu_b^{1/2}$. It follows that the best estimates of μ_a and its standard deviation are $\mu_{ab} - \mu_b$ and $(\mu_{ab} + \mu_b)^{1/2}$, respectively.

Modeling a Poisson process: It is instructive to model a Poisson process on a computer. Step through a series of increments of equal size (period of time, area, volume). Assume a constant probability, λ, that an event will occur in a given increment: get a random number uniformly distributed between 0 and 1; if this number is less than or equal to λ, score a success; if it is greater than λ, score a failure. For radioactive decay, success is a count; failure is no count. For sampling bacteria, success is the presence of a bacterium; failure is its absence. For motility of *E. coli*, success is a transition from the run mode to the tumble mode or back again; failure is its absence.

A reasonably accurate simulation of the motion of *E. coli*, which includes both the effects of rotational diffusion occurring during runs and the highly erratic movement occurring during tumbles, is shown in Fig. 6.11. This particular track is called "A Flamenco Dancer." The occurrence of a track that resembles anything at all, let alone anything animate, is extremely rare.

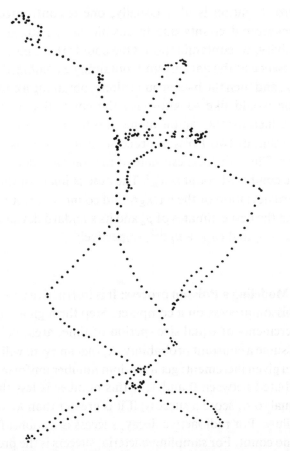

Fig. 6.11. A computer simulation of the motion of *E. coli*. The plot is a projection on the *xy* plane of a three-dimensional random walk. The track began in the run mode at the dancer's lowered hand and ended, on depression of the computer's stop button, in the tumble mode at the upraised hand. In the run mode, each step (the displacement from one dot to the next) is a vector 2 units long making an angle of $\theta = 8°$ with the preceding vector but rotated around it by a random angle between 0 and 360°; as in Fig. 6.8. After each step there is a probability 0.1 that the cell reverts to the tumble mode. In the tumble mode, each step is a vector 1 unit long making an angle of 80° with the preceding vector; after the step there is a probability 0.3 that the cell reverts to the run mode. This track was generated by E. M. Purcell on a desk-top computer in the basement of the Institute for Advanced Study at Princeton, May 1975.

Bacterial diffusion

If a cell swims at a constant speed v along a trajectory comprising a sequence of exponentially distributed straight runs of mean duration τ, it can be shown that its diffusion coefficient is

$$D = \frac{v^2 \tau}{3(1 - \alpha)}, \qquad (6.11)$$

where α is the mean value of the cosine of the angle between successive runs; see Lovely and Dahlquist (1975). If the change in direction from run to run is random, Eq. 6.7 implies $\alpha = 0$, and $D = v^2\tau/3$.

To see this from first principles, go back to Eq. 1.8 and note that $\langle x^2(n) \rangle = n \langle \delta^2 \rangle$, where now δ is the x component of the run length, l. From geometry, $\langle \delta^2 \rangle = \langle l^2 \rangle / 3$. From the fact that the run lengths are exponentially distributed, Eq. 6.9, $\langle l^2 \rangle = 2\langle l \rangle^2$. Since $l = v\tau$ and $n = t/\tau$, we get $\langle x^2(n) \rangle = 2(v^2\tau/3)t$, which, on comparison with Eq. 1.10, gives $D = v^2\tau/3$.

The mean cosine is positive if there is a forward directional bias; this is equivalent to making the runs longer. The mean cosine is negative if there is a backward directional bias; this is equivalent to making the runs shorter. For *E. coli*, $\alpha \simeq 0.33$. With $v \simeq 3 \times 10^{-3}$ cm/sec and $\tau \simeq 1$ sec, we get $D \simeq 4 \times 10^{-6}$ cm²/sec. Thus, this bacterium moves with a diffusion coefficient nearly as large as that of a molecule of low molecular weight, e.g., sucrose. A mutant that tumbles incessantly has a small diffusion coefficient, because τ is smaller. A dead cell or a mutant that is paralyzed has a much smaller diffusion coefficient: approximately that of a sphere of radius 10^{-4} cm, about 2×10^{-9} cm²/sec, as we computed following Eq. 4.13. One might suppose that a mutant that swims but never tumbles would have a very large diffusion coefficient. In this case, the assumption that runs are straight breaks down, and diffusion is limited by rotational Brownian movement. To see this, decompose the trajectory into a sequence of exponentially distributed straight segments of mean duration τ'. If τ' is small, the angle between successive segments, θ, is small, and its cosine is approximately $1 - \theta^2/2$. The mean value of the cosine is $\alpha = 1 - \langle\theta^2\rangle/2$. From Eq. 6.2 and the discussion following Eq. 6.6, we find that $\langle\theta^2\rangle = 4D_r\tau'$, where D_r is the rotational diffusion coefficient of the cell. Therefore, $(1 - \alpha) = 2D_r\tau'$, and Eq. 6.11 gives $D = v^2/6D_r$. For $v \simeq 3 \times 10^{-3}$ cm/sec and $D_r \simeq 0.062$ radians^2/sec, $D \simeq 2 \times 10^{-5}$ cm^2/sec.

If a cell can diffuse this well by working at the limit imposed by rotational Brownian movement, why does it bother to tumble? The answer is that the tumble provides the cell with a mechanism for biasing its random walk. When it swims in a spatial gradient of a chemical attractant or repellent and happens to run in a favorable direction, the probability of tumbling is reduced. As a result, favorable runs are extended, and the cell diffuses with drift. This drift is not imposed by an externally applied force, as was the case for the inert particles discussed in Chapter 4, but rather by an externally applied sensory cue. The cell analyzes this cue and generates the bias internally, by changing the way in which it rotates its flagella. The drift velocity can be enormous, as large as $v/10$. A cell with a drift velocity of this magnitude can easily outdistance one that swims only at random. Given enough time, it can do so with a drift velocity that is very much smaller.

[See further material in Appendix C.]

Chapter 7

Other Random Walks

In Chapter 4 we discussed methods for separating particles that involved drift in externally applied gravitational, centrifugal, or electrical fields. In these methods, the resolution of components of different effective mass or charge is limited by diffusion in the direction of the externally applied field, i.e., by spreading in a direction parallel to the axis along which the separation occurs. Here we describe methods for separating particles in which random motion in a direction normal to this axis plays an essential role.

Countercurrent distribution

If a chemical is mixed with aliquots of two immiscible liquids of volumes V_a and V_b, and the aliquots are shaken together and allowed to stand, then the final concentrations of the chemical in the two liquids, C_a and C_b, will depend upon the relative affinity of the chemical for the two liquids. The probability that a given molecule of a chemical will be found in phase a is equal to the average number of molecules of the chemical in phase a divided by the total number of molecules in the system; i.e.,

$$p = \frac{C_a V_a}{C_a V_a + C_b V_b}. \tag{7.1}$$

The probability that the molecule will be found in phase b is

$$q = \frac{C_b V_b}{C_a V_a + C_b V_b}. \tag{7.2}$$

These events are mutually exclusive; $q = 1 - p$. If the solutions are dilute, p and q depend not on the absolute values of C_a and C_b but only on their ratio, C_a/C_b, which is called the partition coefficient. If the chemical is more soluble in liquid a than in liquid b, the partition coefficient is greater than 1. If, in addition, $V_a \simeq V_b$, then p is greater than 1/2.

This partition experiment can be repeated in an interesting way with a countercurrent apparatus of the sort shown in Fig. 7.1. The chemical (or chemical mixture) is added at position 0, the apparatus is shaken so that the liquids in the paired tubes equilibrate, the phases are allowed to separate, and the upper set of tubes is shifted one position to the right. The process is repeated until the tubes have been shaken and shifted n times. If a given molecule chooses the upper phase k times, it will end up at position k. The total number of trials is n, and the prob-

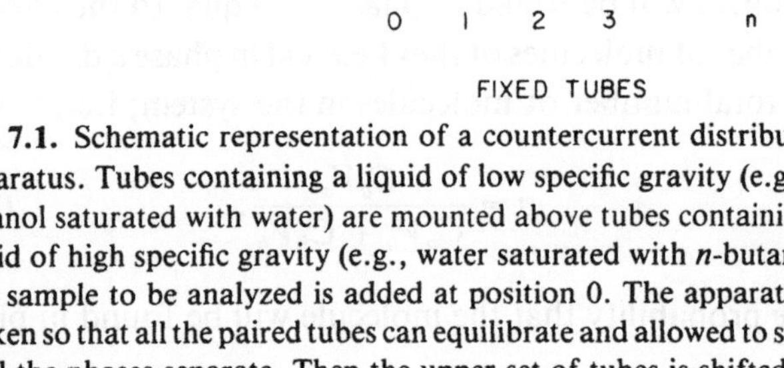

Fig. 7.1. Schematic representation of a countercurrent distribution apparatus. Tubes containing a liquid of low specific gravity (e.g., n-butanol saturated with water) are mounted above tubes containing a liquid of high specific gravity (e.g., water saturated with n-butanol). The sample to be analyzed is added at position 0. The apparatus is shaken so that all the paired tubes can equilibrate and allowed to stand until the phases separate. Then the upper set of tubes is shifted one position to the right, as shown by the arrow, and the process is repeated.

ability that the molecule chooses the upper phase at each trial is p. In effect, the molecule flips a biased coin n times: whenever the result is heads, it moves one position to the right; whenever it is tails, it stays where it is. Thus, the molecule moves in accord with the binomial distribution. The probability that it will be found at position k is $P(k;n,p)$; see Eq. A.18. Molecules of another kind, having, say, a larger partition coefficient, will be distributed in accord with the distribution $P(k;n,p')$, with $p' > p$. Two such distributions with means np and np' and standard deviations $(npq)^{1/2}$ and $(np'q')^{1/2}$ are shown in Fig. 7.2.

If the total number of molecules of each kind is roughly the same, the two peaks will begin to separate when the difference in the means is comparable to the sums of the standard deviations, i.e., when

$$np' - np = (np'q')^{1/2} + (npq)^{1/2}. \qquad (7.3)$$

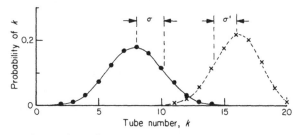

Fig. 7.2. Separation of two chemicals in an apparatus of the sort shown in Fig. 7.1 after $n = 20$ transfers. The probability of choosing the upper phase is $p = 0.4$ for one chemical (●), and $p' = 0.8$ for the other (x). The means and standard deviations of the two distributions are $\langle k \rangle = 8$, $\sigma = (npq)^{1/2} = 2.2$, and $\langle k' \rangle = 16$, $\sigma' = (np'q')^{1/2} = 1.8$, respectively. The two distributions overlap, but the probability that a molecule of the first kind will be found in tube 16 is only 3×10^{-4}, and the probability that a molecule of the second kind will be found in tube 8 is only 9×10^{-5}.

If we double n, the means increase by a factor of 2, the standard deviations by a factor of $2^{1/2}$, so the separation improves by a factor of $2^{1/2}$. The criterion for reasonably good separation is

$$n^{1/2}(p' - p) > (p'q')^{1/2} + (pq)^{1/2}. \qquad (7.4)$$

Note that a separation always can be effected if n is large enough, provided that $p' \neq p$.

Velocity of a molecule down the countercurrent apparatus: It is instructive to think about countercurrent distribution from a slightly different point of view. Suppose that we plot the *velocity* of a molecule of a given kind down the apparatus of Fig. 7.1, assuming that the tubes are a distance δ apart and that a transfer is made every τ seconds at an average velocity $v_0 = \delta/\tau$. Plots of this kind for two such molecules are shown in Fig. 7.3, for the case $n = 10$. The first molecule chose the upper phase before the 2nd, 3rd, 6th, 7th and, 9th transfers; the second molecule chose the upper phase before the 1st, 2nd, 4th, and

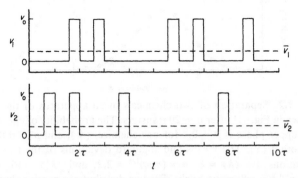

Fig. 7.3. The velocities of two identical molecules down the apparatus of Fig. 7.1. In this experiment, the upper set of tubes was moved one position to the right at an average velocity v_0 every τ seconds for a period of time $t_0 = 10\tau$. The average velocities of the molecules over this period are denoted $\bar{v}_1$ and $\bar{v}_2$ (dashed lines).

8th transfers. The average velocity of the first molecule over the time period 10τ was $\bar{v}_1$; the average velocity of the second molecule was $\bar{v}_2$; evidently, $\bar{v}_2 = (4/5)\bar{v}_1$. The final distribution of molecules will be narrower the more closely these average velocities agree with one another.

The number of bumps in a velocity curve, Fig. 7.3, equals k, the number of times the molecule shifts one position to the right. The mean velocity of a molecule is its displacement, $k\delta$, divided by the duration of the experiment, t_0:

$$\bar{v} = k\delta/t_0. \qquad (7.5)$$

The mean velocities, $\bar{v}$, for different molecules are distributed in the same way as the final positions, $k\delta$. In particular,

$$\langle \bar{v} \rangle = \langle k \rangle \delta/t_0 = np\delta/t_0, \qquad (7.6)$$

and

$$\sigma_v = [\langle \bar{v}^2 \rangle - \langle \bar{v} \rangle^2]^{1/2} = \sigma_k \delta/t_0 = (npq)^{1/2} \delta/t_0; \qquad (7.7)$$

see Eqs. A.22 and A.25. Note that the bar in these expressions denotes an average for a single molecule over the period of time t_0, while the brackets denote an average over the entire ensemble of molecules. The fractional width of the distribution is

$$\frac{\sigma_v}{\langle \bar{v} \rangle} = \frac{(npq)^{1/2}}{np} = \left(\frac{q}{np}\right)^{1/2}. \qquad (7.8)$$

The fractional width decreases as $(np)^{1/2} = \langle k \rangle^{1/2}$, i.e., as the square-root of the average number of bumps in the velocity curves. The larger the number of bumps, the better the separation. Or to put it another way, the larger the number of times a molecule changes phases, the more likely its history will resemble that of other molecules of the same kind, i.e., the more likely the molecules will move together as a group. Separations depend not only on the fact that different kinds of molecules have different affinities for the two phases but also on the fact that molecules of each kind repeatedly move from one phase to the other.

Partition chromatography

Statistical averaging is fundamental to all of chromatography. Instead of a set of fixed tubes containing a liquid b, we use a fixed solid, such as paper, a layer of silica gel, an ion-exchange resin, or porous beads made of cross-linked polyacrylamide, dextran, glass, etc. Instead of a set of moveable tubes containing a liquid a, we use a moving stream of solvent, such as butanol saturated with water, a mixture of isopropanol, acetic acid and water, a dilute aqueous buffer, etc. The velocity of the ith molecule of a given kind down the apparatus now looks something like the plot shown in Fig. 7.4, a curve that is not as regular as those shown in Fig. 7.3 but one that has the same essential features. Here the number of bumps depends on the relative affinities of the molecule for the fixed and moving phases and on the rate at which the molecule can diffuse from one to the other. One needn't shake the apparatus and shift tubes from one position to the next; the molecules wander back and forth from the fixed to the moving phases on their own. While they are in the moving phase, they drift with the flow and move down the apparatus. The details depend on adsorption and diffusion times, flow rates, and the like, but never mind. The important thing is that the separation improves as the square-root of the number of bumps in the velocity curves, i.e., as the square root of the number of transits from phase to phase. Since these increase linearly with

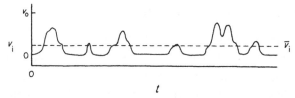

Fig. 7.4. The drift velocity of molecule i during partition chromatography. The molecule made about six transits from the fixed to the moving phase. Its average velocity was $\bar{v}_i$. An example might be leucine in a paper chromatogram eluted with butanol saturated with water: if so, periods of 0 velocity occur during partition into water held fixed by the fibers of cellulose in the paper, and periods of finite velocity occur during partition into butanol moving down the paper between these fibers. Another example might be lysozyme in a column packed with porous dextran beads eluted with a dilute aqueous buffer: if so, periods of 0 velocity occur during partition into buffer fixed within the pores of the beads, and periods of finite velocity occur during partition into buffer moving down the column between these beads. In the first example, the separation is on the basis of chemical affinity (preference for water); in the second, the separation is on the basis of size (ability to diffuse into the pores in the beads).

time, the separation improves with the square-root of the time. One needn't lengthen the apparatus; it is sufficient to decrease the rate of flow. The highest resolution is obtained with schemes that provide the largest number of transits from phase to phase, such as gas chromatography, where diffusion coefficients are large, or high-pressure liquid chromatography, where spacings between the particles that compose the fixed phase are very small.

In these experiments, it also is possible for molecules to diffuse in a direction parallel to the axis along which the separation occurs, i.e., with or against the flow. In practice, however, the duration of the experiment is rarely long enough for spreading due to this mechanism to be significant. There are more serious problems that have to do with nonuniformities of rates of flow from one side of the paper or column to the other.

Sedimentation field-flow fractionation

There is one variant of chromatography in which the affinity for the fixed, or more slowly moving, phase is determined by adherence to the Boltzmann law, Eq. 5.4. In this variant, the statistical behavior of the particles is completely determined, and separation can be made on the basis of effective mass. A schematic representation of the apparatus is shown in Fig. 7.5. As explained in the figure legend, particles to be analyzed are placed at one end of a long channel and allowed to sediment to equilibrium. Fluid injected at this end of the apparatus moves slowly down the channel with a parabolic velocity profile determined by viscous flow, Eq. 4.9. Near the bottom of the channel the profile is nearly linear: the higher up a particle goes, the faster it drifts down the apparatus. Thus,

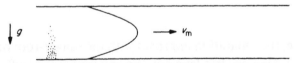

Fig. 7.5. A schematic representation of a sedimentation field-flow fractionation apparatus. A channel is filled with a fluid, e.g., a dilute aqueous buffer, and exposed to an externally applied field, g. Particles are suspended at the left end of the channel and allowed to settle to equilibrium. Then fluid is injected at a constant rate at the left end of the channel and removed at the right end. It moves slowly through the channel with a parabolic velocity profile, with a maximum velocity, v_m, at the center of the channel. Fractions of fluid emerging at the right end of the channel are collected and analyzed for particles. Lighter particles spend more time in regions of the fluid that flow more rapidly and are eluted first. Heavier particles are eluted later. The theory of the method is given in Berg and Purcell (1967a), an apparatus utilizing a gravitational field is described in Berg, Stewart, and Purcell (1967), and an apparatus utilizing a centrifugal field is described in Berg and Purcell (1967b). For recent applications, see Kirkland and Yau (1982), Giddings (1988), and Berg and Turner (1991).

particles of smaller effective mass, which spend more time on the average higher up, leave the apparatus sooner than particles of larger effective mass. But time must be allowed for the particles to diffuse up and down across their Boltzmann distributions, i.e., to move back and forth between the more slowly and more rapidly moving portions of the fluid, so that the average histories of particles of a given kind are nearly the same. If such time is not allowed, particles that begin the game near the top of their Boltzmann distribution will leave the apparatus much sooner than particles of the same kind that begin the game near the bottom of their Boltzmann distribution. Since the time for diffusion across the Boltzmann distribution depends on the size of the diffusion coefficient of the particle, the width of the emerging band provides a measure of this coefficient. Thus, in principle, one can separate a mixture of particles according to their effective masses and measure the effective mass and diffusion coefficient of each kind of particle, all in one experiment.

The Boltzmann distribution, Eq. 5.4, is

$$\frac{C(x)}{C(0)} = e^{-x/x_s}, \qquad (7.9)$$

where $C(x)$ and $C(0)$ are the concentrations of particles at height x and 0, respectively, and x_s is the scale height,

$$x_s = \frac{kT}{m'g}. \qquad (7.10)$$

Here m' is the effective mass; see Eq. 4.17. The time required to diffuse a root-mean-square distance x_s is, by Eq. 1.10,

$$t_D = \frac{x_s^2}{2D}. \qquad (7.11)$$

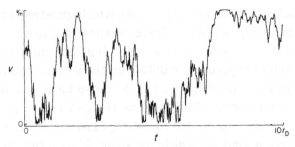

Fig. 7.6. The velocity of a single particle down the channel during sedimentation field-flow fractionation as a function of time. The scale height of the particle is $x_s = 0.2x_0$, where x_0 is the height of the channel. The motion of the particle was followed for a period of time $t = 10t_D$, where t_D is the time for diffusion a root-mean-square distance x_s, Eq. 7.11. The simulation was done on a computer by starting the particle at height x_s, allowing it to diffuse up or down a distance $x_0/200$, assigning to it a velocity equal to the velocity of the fluid at the new height, and repeating the process 16,000 times. Sedimentation was accounted for by making the probability of diffusing up smaller than the probability of diffusing down by a factor $e^{-1/40}$, where 40 is the scale height in units of $x_0/200$. At the bottom of the channel the particle only was allowed to diffuse up; at the top of the channel it only was allowed to diffuse down. The factor $e^{-1/40}$ ensures that the particle will be found at height x with probability e^{-x/x_s}, as required by the Boltzmann law, Eq. 7.9.

Thus, the average number of transits from the more slowly to the more rapidly moving portions of the fluid that occur during the time, t_0, that the particle spends in the apparatus is of order t_0/t_D. As we have seen in Eq. 7.8, the resolution should improve as the square-root of this number, namely, as $(t_0/t_D)^{1/2}$.

A computer simulation of the drift velocity of a molecule having a scale height equal to 1/5 of the channel height is shown in Fig. 7.6. Elution profiles expected for particles of this kind in a series of experiments in which the flow rates are successively halved are shown in Fig. 7.7.

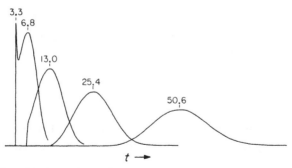

Fig. 7.7. Elution of equal numbers of identical particles from a sedimentation field-flow fractionation apparatus. A series of four simulations is shown, in which the average velocity of the fluid was successively halved. In each simulation, the flow began at the left end of the abscissa. The numbers over the peaks refer to the times that particles which emerged at that point spent in the channel, in units of the diffusion time, t_D. Other conditions are the same as in Fig. 7.6. In the first simulation, a secondary peak appeared at $t \simeq 3.3t_D$. This peak contains particles moving at $v \simeq v_m$ that did not have time to diffuse away from the center of the channel, where the flow velocity changes relatively little with height; see Fig. 7.5. In time $t \simeq 3.3t_D$, the particles diffuse a root-mean-square distance $1.8x_s = 0.36x_0$, i.e., less than half the height of the gap. The secondary peak disappeared for times of elution of order $13t_D$. After this, the peaks were nearly symmetric. When the flow rate is halved, the particles emerge over a period of time that increases by a factor of $2^{1/2}$, but the mean time of emergence increases by a factor of 2, so the resolution improves by a factor of $2^{1/2}$.

Autocorrelation analysis

As noted in the discussion of Eq. 7.8, the ability of a chromatographic procedure to distinguish particles of different kinds depends on the narrowness of distributions of mean velocities

$$\bar{v} = \frac{1}{t_0} \int_0^{t_0} v(t)\, dt, \qquad (7.12)$$

where $v(t)$ is the velocity in the direction of separation of a particular molecule, and the average is over the duration of the experiment, t_0. The fractional width of the distribution for particles of one kind, $\sigma_v/\langle \bar{v} \rangle$, depends on the mean $\langle \bar{v} \rangle$ and the variance

$$\sigma_v^2 = \langle \bar{v}^2 \rangle - \langle \bar{v} \rangle^2, \qquad (7.13)$$

where the averages denoted by the brackets are over the ensemble of N particles. A useful technique for analyzing this problem involves use of the autocorrelation function

$$G(\tau) = \langle v(t)v(t+\tau) \rangle, \qquad (7.14)$$

the ensemble mean of the product of $v(t)$ and $v(t+\tau)$. From the definition of $\bar{v}$, Eq. 7.12, it follows that

$$\bar{v}^2 = \frac{1}{t_0^2} \int_0^{t_0} dt' \int_0^{t_0} v(t')v(t)\,dt. \qquad (7.15)$$

By introducing the autocorrelation function, we obtain

$$\langle \bar{v}^2 \rangle = \frac{1}{t_0^2} \int_0^{t_0} dt' \int_0^{t_0} G(t'-t)\,dt. \qquad (7.16)$$

This reduces the problem of determining $\langle \bar{v}^2 \rangle$ and σ_v^2 to one of determining $G(\tau)$.

Autocorrelation analysis of partition chromatography: As an example of how such a calculation can be made, consider a simplified model of partition chromatography in which the velocity of a particle is 1 when it is free, 0 when it is fixed. Let the probability per unit time that a particle leaves the moving phase be λ_1 and the probability per unit time that it returns to the moving phase be λ_0. With these assumptions, the mean dwell-time in the moving phase is $1/\lambda_1$ and the mean dwell-time in the fixed phase is $1/\lambda_0$; see the discussion of the Poisson interval distribution given in Chapter 6. The mean velocity is equal to the fraction of time that the particle spends in the moving phase:

$$\langle \bar{v} \rangle = \frac{1/\lambda_1}{1/\lambda_1 + 1/\lambda_0} = \frac{\lambda_0}{\lambda_1 + \lambda_0}. \quad (7.17)$$

To compute the autocorrelation function, consider the ensemble of N particles at an arbitrary time t. Recall that $v(t)$ is either 1 or 0. If N is large, about $N\langle \bar{v} \rangle$ of the particles will have velocity $v(t) = 1$. By the definition of $G(\tau)$, about $NG(\tau)$ of these will have velocity $v(t + \tau) = 1$; these "1,1" particles are the only ones for which $v(t)v(t + \tau) \neq 0$. Now consider the result of shifting the time of the second observation from $t + \tau$ to $t + \tau + d\tau$. Some of the $NG(\tau)$ 1,1 particles will become 1,0 particles and some of the $N\langle \bar{v} \rangle - NG(\tau)$ 1,0 particles will become 1,1 particles. Since the probability that a particle leaves the moving phase in time $d\tau$ is $\lambda_1 d\tau$, $NG(\tau)\lambda_1 d\tau$ of the 1,1 particles will become 1,0 particles. Since the probability that a particle leaves the fixed phase in time $d\tau$ is $\lambda_0 d\tau$, $[N\langle \bar{v} \rangle - NG(\tau)]\lambda_0 d\tau$ of the 1,0 particles will become 1,1 particles. The total number of 1,1 particles will change from $NG(\tau)$ to $NG(\tau + d\tau)$. This allows us to write a difference equation for $G(\tau)$. The net change in the number of 1,1 particles is

$$NG(\tau + d\tau) - NG(\tau) = -NG(\tau)\lambda_1 d\tau$$
$$+ [N\langle \bar{v} \rangle - NG(\tau)]\lambda_0 d\tau. \quad (7.18)$$

This implies that

$$\frac{dG(\tau)}{d\tau} = \langle \bar{v} \rangle \lambda_0 - (\lambda_1 + \lambda_0)G(\tau). \quad (7.19)$$

Note from Eq. 7.17 that $\langle \bar{v} \rangle \lambda_0 = \langle \bar{v} \rangle^2 (\lambda_1 + \lambda_0)$. The value of $G(0)$ is $\langle v^2(t) \rangle$, but since $v(t)$ is either 1 or 0, $\langle v^2(t) \rangle = \langle v(t) \rangle = \langle \bar{v} \rangle$. If τ is large, there will be no correlation between the number of particles for which

108—Other Random Walks

$v(t) = 1$ and $v(t + \tau) = 1$, and $G(\tau)$ must approach $\langle \bar{v} \rangle^2$. All these criteria are met by the solution

$$G(\tau) = \langle \bar{v} \rangle^2 + \langle \bar{v} \rangle (1 - \langle \bar{v} \rangle) e^{-(\lambda_1 + \lambda_0)|\tau|}, \quad (7.20)$$

where $|\tau|$ is the absolute value of τ. This function is shown in Fig. 7.8

We now use Eqs. 7.16 and 7.13 to compute σ_v^2. In the limit $t_0 \gg 1/(\lambda_1 + \lambda_0)$, we obtain

$$\sigma_v^2 = \frac{2 \langle \bar{v} \rangle (1 - \langle \bar{v} \rangle)}{(\lambda_1 + \lambda_0) t_0}. \quad (7.21)$$

Given Eq. 7.17, this can be written

$$\frac{\sigma_v}{\langle \bar{v} \rangle} = \frac{2^{1/2} \lambda_1/(\lambda_1 + \lambda_0)}{[t_0/(1/\lambda_1 + 1/\lambda_0)]^{1/2}}. \quad (7.22)$$

Apart from the square-root of 2, this equation states that the fractional width of the velocity distribution is equal to the fraction of time that a molecule spends in the fixed phase divided by the square-root of the number of excursions that it makes into the moving phase during the run-

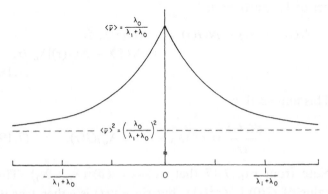

Fig. 7.8. Autocorrelation function for particles that change their velocities at random from 0 to 1 and 1 to 0 with probabilities per unit time λ_0 and λ_1, respectively. See Eq. 7.20.

ning time, t_0. Once again, we see that the resolution improves as the square-root of the number of bumps in the velocity curves.

The autocorrelation function provides a measure of the interval of time, called the correlation time, over which the behavior of a particle depends on its recent past. If a particle is in one phase at time t, it is likely to be found in the same phase at time $t + \tau$, provided that $\tau < 1/(\lambda_1 + \lambda_0)$. If $\tau \gg 1/(\lambda_1 + \lambda_0)$, there is no way of predicting where the particle will be, apart from saying that it will be found in the moving phase with probability $\lambda_0/(\lambda_1 + \lambda_0)$.

Autocorrelation analyses of other processes: Similar calculations can be made for any process in which transitions occur between two states at random, e.g., for transitions between run and tumble modes of a swimming bacterium, between open and closed configurations of a membrane channel, or between bound and free states of a specific receptor. The latter problem is discussed in Berg and Purcell (1977). The calculation of the autocorrelation function given above was adapted from that source.

In many processes of interest, called stationary random processes, the behavior of N particles over an interval of time t_0 is statistically equivalent to the behavior of a single particle over the much longer interval of time Nt_0. The statistical analysis proceeds as before, but the averages denoted by the bar in Eqs. 7.12 et. seq. can refer to the behavior of one particle over successive intervals of time, t_0; and the averages denoted by the brackets in Eqs. 7.13 et. seq. can refer to the behavior of the same particle over an unlimited period of time. This situation occurs whenever a recording is made of a signal, $v(t)$, by an instrument with a finite resolving time, t_0. The instrument does not record the instantaneous value of the function $v(t)$

but rather successive average, or running average, values $\bar{v}$. From measurements of the variance of this output, however, one can deduce some of the microscopic properties of the process that generates $v(t)$, as evidenced, for example, by the appearance in Eq. 7.21 of the transition probabilities λ_1 and λ_0.

Finally, it should be mentioned that complete information about the distribution in frequency of the power in the signal $v(t)$ is contained in its autocorrelation function. The power spectral density and the autocorrelation function are Fourier transform pairs. An equivalent analysis of the statistical behavior of $v(t)$ can be made in the frequency domain. See, for example, Davenport and Root (1958).

Appendix A

Probabilities and Probability Distributions

Here is a primer on the theory of probability for those who are new to the subject. The concepts are deceptively simple: easy to define but not easily made part of one's working vocabulary.

Probabilities

Classical or *a priori* probabilities are defined in terms of the possible outcomes of a trial, recognized in advance as equally probable. In the toss of a coin, the probability of getting a head is 1/2: the number of outcomes that give a head, 1, divided by the total number of possible outcomes, head or tail, 2. In the toss of a die, the probability of getting one dot is 1/6: the number of outcomes that give one dot, 1, divided by the total number of possible outcomes, one through six dots, 6. In general, the probability of event a is

$$p(a) = \frac{a}{n}. \tag{A.1}$$

where a is the number of equally probable outcomes that satisfy criteria a, and n is the total number of equally probable outcomes.

In the examples just given, the outcomes are *mutually exclusive*; i.e., only one outcome is possible at a time. If events a and b are mutually exclusive, then

$$p(a \text{ or } b) = p(a) + p(b), \tag{A.2}$$

and

$$p(\text{not } a) = 1 - p(a). \tag{A.3}$$

For the toss of a coin, p(head or tail) = p(head) + p(tail) = 1, and p(not head) = 1 − p(head) = 1/2. For the toss of a die, p(1 dot or 2 dots) = p(1 dot) + p(2 dots) = 1/3, and p(not 1 dot) = 1 − p(1 dot) = 5/6.

In these examples, the outcomes also are *statistically independent*; i.e., the occurrence of one event does not affect that of another. If events a and b are statistically independent, then

$$p(a \text{ and } b) = p(a)p(b). \qquad (A.4)$$

The probability of obtaining heads in each of two tosses of a coin is (1/2)(1/2) = 1/4. The probability of obtaining a single dot in each of two tosses of a die is (1/6)(1/6) = 1/36.

Events are *conditional* if the probability of one event depends on the occurrence of another. If the probability that b will occur, given that a has occurred, is $p(b/a)$, then the probability that both will occur is

$$p(a \text{ and } b) = p(a)p(b/a). \qquad (A.5)$$

For example, the probability of drawing two aces from a deck of cards is (4/52)(3/51) = 1/221. If the first card were put back into the deck and the deck reshuffled, then the probability of drawing the second ace would not be conditioned on the drawing of the first, and the probability would be (4/52)(4/52) = 1/169, in accord with Eq. A.4.

Here are some rules of thumb for dealing with *compound events*, when the events are both mutually exclusive and statistically independent:

$$p(\text{neither } a \text{ nor } b) = p(\text{not } a \text{ and not } b)$$
$$= [1 - p(a)][1 - p(b)] \qquad (A.6)$$

$$p(\text{either } a \text{ or } b, \text{ but not both}) = p(a \text{ and not } b, \text{ or not } a \text{ and } b)$$
$$= p(a)[1 - p(b)] + [1 - p(a)]p(b) \quad (A.7)$$

$$p(\text{either } a \text{ or } b \text{ or both}) = 1 - p(\text{neither } a \text{ nor } b). \quad (A.8)$$

Probability distributions

Suppose we were to toss an unbiased coin 4 times in succession. What is the probability, $P(k)$, of obtaining k heads? There are 16 different ways the coins might land; each is equally probable. Let's write down all 16 but group them according to how many heads appear, using the binary notation 1 = heads, 0 = tails:

Sequence	No. heads	$P(k)$
0000	0	$P(0) = 1/16$
1000	1	
0100	1	$P(1) = 4/16 = 1/4$
0010	1	
0001	1	
1100	2	
1010	2	
1001	2	$P(2) = 6/16 = 3/8$
0110	2	
0101	2	
0011	2	
1110	3	
1101	3	$P(3) = 4/16 = 1/4$
1011	3	
0111	3	
1111	4	$P(4) = 1/16$

From Eq. A.1, the probability of a given sequence is 1/16. From Eq. A.2, the probability of obtaining the sequence 1000, 0100, 0010, or 0001, i.e., a sequence in which 1 head appears, is $1/16 + 1/16 + 1/16 + 1/16 = 1/4$. The third column lists these probabilities, $P(k)$.

The results of these calculations can be summarized by plotting $P(k)$ as a function of k, as shown in Fig. A.1. Such a plot is called a theoretical probability distribution. The peak value, $k = 2$, is the *most probable* value. Since the curve is symmetric about $k = 2$, this value also must be the mean value. One can compute the *mean* or *expectation value* of k by adding the number of heads obtained for each of the sequences shown in the table and dividing by 16, or—and this is the same thing—by weighting each possible value of k by the probability of obtaining that

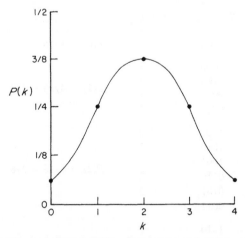

Fig. A.1. The probability, $P(k)$, of obtaining k heads in 4 tosses of an unbiased coin. This theoretical probability distribution is *discrete*; it is defined only for the integer values $k = 0, 1, 2, 3$, or 4. A smooth line is drawn through the points only to make it easier to visualize the trend. The distribution has a mean value 2 and a standard deviation 1.

value, $P(k)$, and computing the sum:

$$\langle k \rangle = \sum_{k=0}^{4} kP(k). \quad (A.9)$$

The value of this sum is $(0)(1/16) + (1)(1/4) + (2)(3/8) + (3)(1/4) + (4)(1/16) = 0 + 1/4 + 3/4 + 3/4 + 1/4 = 2$, as expected. Note that

$$\sum_{k=0}^{4} P(k) = 1. \quad (A.10)$$

The probability of obtaining 0, 1, 2, 3, or 4 heads is 1. The distribution is properly *normalized*.

Note also that if a is a constant (does not depend on the variable k),

$$\langle a \rangle = \sum_{k=0}^{4} aP(k) = a \sum_{k=0}^{4} P(k) = a, \quad (A.11)$$

and

$$\langle ak \rangle = \sum_{k=0}^{4} akP(k) = a \sum_{k=0}^{4} kP(k) = a\langle k \rangle. \quad (A.12)$$

It is useful to have some measure of the width or spread of a distribution about its mean. One might compute $\langle k - \mu \rangle$, the expectation value of the deviation of k from the mean $\mu = \langle k \rangle$, but the answer always comes out 0. It makes more sense to compute the expection value of the square of the deviation, namely

$$\sigma^2 = \langle (k - \mu)^2 \rangle = \langle k^2 - 2\mu k + \mu^2 \rangle. \quad (A.13)$$

This quantity is called the *variance*. Its square root, σ, is called the *standard deviation*. Since $\mu = \langle k \rangle$ and $\mu^2 = \langle k \rangle^2$ are constants, Eq. A.13 can be simplified. It follows from Eqs. A.11 and A.12 that

$$\sigma^2 = \langle k^2 \rangle - 2\mu\langle k \rangle + \mu^2 = \langle k^2 \rangle - \langle k \rangle^2, \quad (A.14)$$

where

$$\langle k^2 \rangle = \sum_{k=0}^{4} k^2 P(k). \qquad (A.15)$$

For the distribution of Fig. A.1, $\langle k^2 \rangle = (0)(1/16) + (1)(1/4) + (4)(3/8) + (9)(1/4) + (16)(1/16) = 0 + 1/4 + 6/4 + 9/4 + 1 = 5$, $\langle k \rangle^2 = 4$, and $\sigma^2 = 5 - 4 = 1$.

It is instructive to sit down and actually flip a coin 4 times in succession, count the number of heads, and then repeat the experiment a large number of times. One can then construct an *experimental* probability distribution, with $P(0)$ equal to the number of experiments that give 0 heads divided by the total number of experiments, $P(1)$ equal to the number of experiments that give 1 head divided by the total number of experiments, etc. Two such distributions are shown in Fig. A.2. In the first (x),

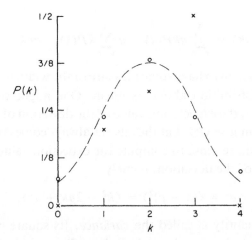

Fig. A.2. Two experimental probability distributions. In one (x), a coin was flipped 4 times in 10 successive experiments; the mean value is 2.30, the standard deviation is 1.22. In the other (o), a coin was flipped 4 times in 100 successive experiments; the mean value is 2.04, the standard deviation 1.05. The dashed curve is the theoretical probability distribution of Fig. A.1.

the total number of experiments was 10. In the second (o), the total number was 100.

If you do not like flipping coins, you can build a probability machine of the sort shown in Fig. A.3. If the machine is level and well made, the probability that a ball bounces to the right or the left on striking the next pin is 1/2. The number of successive trials is equal to the number of rows of pins. If you drop 100 balls through this machine, they will pile up in the bins at the bottom, forming a distribution like the one shown in Fig. A.2. Or do the experiments on a computer: ask for a random number uniformly distributed between 0 and 1; if the number is less than or equal to 1/2, call it a head; if it is greater than 1/2, call it a tail. The data shown in Fig. A.2 were generated in this way.

The theoretical expectations are more closely met the larger the number of samples. What is the likelihood that the deviations between the experimental distributions and

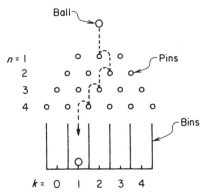

Fig. A.3. A probability machine of 4 rows. A ball dropped on the center pin of the top row bounces to the right or left and executes a random walk as it moves from row to row, arriving finally at one of the bins at the bottom. The rows and bins are numbered $n = 1, 2, 3, 4$ and $k = 0, 1, 2, 3, 4$, respectively; n is the bounce number and k is the number of bounces made to the right.

the theoretical distribution, evident in Fig. A.2, occur by chance? By how much are the mean values of the experimental distributions likely to differ from the mean value predicted by the theory? Questions of this kind often are encountered in the laboratory. They are dealt with in books on data reduction and error analysis and will not be pursued here.

The binomial distribution

What if we flip a biased coin, with the probability of a head p and the probability of a tail $q = 1 - p$? The probability of a given sequence, e.g., 100010 ..., in which k heads appear in n flips is, by Eq. A.4, $pqqqpq$ ..., or

$$p^k q^{n-k}. \tag{A.16}$$

There are a total of 2^n possible sequences. Only some of these give k heads and $n - k$ tails. Their number is

$$\frac{n!}{k!(n-k)!} \equiv \binom{n}{k}, \tag{A.17}$$

where 0!, whenever it appears in the denominator, is understood to be 1.

Since any one or another of these sequences will do, the probability that exactly k heads occur in n flips is, by Eq. A.2.

$$P(k;n,p) = \binom{n}{k} p^k q^{n-k}. \tag{A.18}$$

This is the binomial distribution. The coefficient $\binom{n}{k}$ is the binomial coefficient, the number of combinations of n things taken k and $n - k$ at a time. You have seen it before in algebra in the binomial theorem:

$$(a + b)^n = \sum_{k=0}^{n} \binom{n}{k} a^k b^{n-k}. \tag{A.19}$$

We can use the binomial theorem to show that the binomial distribution is normalized:

$$\sum_{k=0}^{n} P(k;n,p) = \sum_{k=0}^{n} \binom{n}{k} p^k q^{n-k} = (p+q)^n = 1^n = 1.$$

(A.20)

As an example, let's work out the case of 4 flips of an unbiased coin. If $p = q = 1/2$, then $p^k q^{n-k} = (1/2)^n = (1/2)^4 = 1/16$ for all values of k, and the probabilities $P(0;4,1/2), \ldots, P(4;4,1/2)$ are equal to the binomial coefficients $\binom{4}{0}, \ldots, \binom{4}{4}$ times this factor. Since

$$\binom{4}{0} = \binom{4}{4} = \frac{4!}{0!4!} = 1,$$

$$\binom{4}{1} = \binom{4}{3} = \frac{4!}{1!3!} = 4$$

and

$$\binom{4}{2} = \frac{4!}{2!2!} = 6,$$

we obtain the probabilities 1/16, 1/4, 3/8, 1/4, and 1/16, as before.

The expectation value of k is

$$\langle k \rangle = \sum_{k=0}^{n} k P(k;n,p) = \sum_{k=0}^{n} k \frac{n!}{k!(n-k)!} p^k q^{n-k}.$$

(A.21)

To evaluate this, note that the $k = 0$ term is 0 and that $k/k! = 1/(k-1)!$, so that

$$\langle k \rangle = \sum_{k=1}^{n} \frac{n!}{(k-1)!(n-k)!} p^k q^{n-k}.$$

Next, factor out np:

$$\langle k \rangle = np \sum_{k=1}^{n} \frac{(n-1)!}{(k-1)!(n-k)!} p^{k-1} q^{n-k}.$$

Finally, change variables by substituting $m = k - 1$ and $s = n - 1$:

$$\langle k \rangle = np \sum_{m=0}^{s} \frac{s!}{m!(s-m)!} p^m q^{s-m}$$

$$= np \sum_{m=0}^{s} \binom{s}{m} p^m q^{s-m}.$$

The sum in this expression is the same as the one in Eq. A.20; only the labels have been changed. Thus,

$$\langle k \rangle = np. \qquad (A.22)$$

One can evaluate the expectation value of k^2 in a similar fashion by two successive changes in variables and show that

$$\langle k^2 \rangle = (np)^2 + npq. \qquad (A.23)$$

The variance of k, Eq. A.14, is

$$\sigma_k^2 = \langle k^2 \rangle - \langle k \rangle^2 = npq, \qquad (A.24)$$

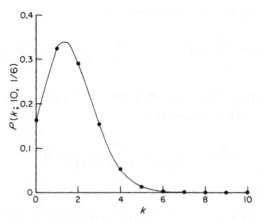

Fig. A.4. The binomial distribution for $n = 10$, $p = 1/6$. The mean value is 1.67, the standard deviation 1.18.

Appendix A—121

and its standard deviation is

$$\sigma_k = (npq)^{1/2}. \quad (A.25)$$

An example of the binomial distribution is given in Fig. A.4, which shows the theoretical distribution $P(k;10,1/6)$. This is the probability of obtaining a given side k times in 10 throws of a die.

The Gaussian distribution

In many of the problems dealt with in this book, the number of trials, n, is very large. A small molecule undergoing diffusion, for example, steps to the right or left millions of times in a microsecond, not 4 times in a few seconds, as the ball in the apparatus of Fig. A.3. There are two asymptotic limits of the binomial distribution. One, the Gaussian, or normal, distribution, is obtained when the probability of a success, p, is finite, i.e., if $np \to \infty$ as $n \to \infty$. The other, the Poisson distribution, is obtained if p is very small, so small that np remains finite as $n \to \infty$.

The derivation of the Gaussian distribution involves the use of Stirling's approximation for the factorials of the binomial coefficients:

$$n! \simeq (2\pi n)^{1/2}(n/e)^n, \quad (A.26)$$

where e is the base of the natural logarithms. The result is

$$P(k;n,p) \to P(k;\mu,\sigma)dk = \frac{1}{(2\pi\sigma^2)^{1/2}} e^{-(k-\mu)^2/2\sigma^2} dk, \quad (A.27)$$

where $\mu = \langle k \rangle = np$ and $\sigma = (\langle k^2 \rangle - \langle k \rangle^2)^{1/2} = (npq)^{1/2}$, as before. $P(k;\mu,\sigma)\,dk$ is the probability that k will be found between k and $k + dk$, where dk is infinitesimal. The distribution is continuous rather than discrete. Expectation values are found by taking integrals rather than sums. The distribution is symmetric about the mean,

μ, and its width is determined by σ. The area of the distribution is 1, so its height is inversely proportional to σ.

If we define $u = (k - \mu)/\sigma$, i.e., plot the distribution with the abscissa in units of σ and the origin at μ, then

$$P(k)\,dk \to P(u)\,du = \frac{1}{(2\pi)^{1/2}} e^{-u^2/2}\,du. \quad (A.28)$$

$P(u)$ is called the normal curve of error; it is shown in Fig. A.5. As an exercise, use your tables of definite integrals and show that

$$\int_{-\infty}^{\infty} P(u)\,du = 1, \quad (A.29)$$

$$\int_{-\infty}^{\infty} uP(u)\,du = 0, \quad (A.30)$$

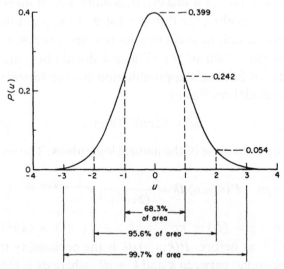

Fig. A.5. The normal curve of error: the Gaussian distribution plotted in units of the standard deviation σ with its origin at the mean value μ. The area under the curve is 1. Half the area falls between $u = \pm 0.67$.

and

$$\int_{-\infty}^{\infty} u^2 P(u)\, du = 1. \tag{A.31}$$

Eq. A.30 can be done by inspection: $P(u)$ is an even function of u, so $uP(u)$ must be an odd function of u. The distribution $P(u)$ is normalized, its mean value is 0, and its variance and standard deviation are 1.

The Poisson distribution

As noted above, the Poisson distribution is obtained as an asymptotic limit of the binomial distribution when p is very small. The result is

$$P(k;\mu) = \frac{\mu^k}{k!} e^{-\mu}, \tag{A.32}$$

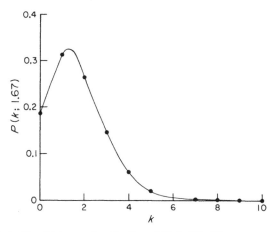

Fig. A.6. The Poisson distribution $P(k;1.67)$. The mean value is 1.67, the standard deviation 1.29. The curve is similar to the binomial distribution shown in Fig. A.4, but it is defined for values of $k > 10$. For example, $P(20;1.67) = 2.2 \times 10^{-15}$.

where $\mu = np$, as before, and 0! is understood to be 1. This distribution is determined by one rather than two constants: $\sigma = (npq)^{1/2}$, but $q = 1 - p \simeq 1$, so $\sigma = (np)^{1/2} = \mu^{1/2}$. The standard deviation is equal to the square-root of the mean. The Poisson distribution is discrete: $P(0;\mu) = e^{-\mu}$ is the probability of 0 successes, given that the mean number of successes is μ, etc. The probability of 1 or more successes is $1 - P(0;\mu) = 1 - e^{-\mu}$. The distribution $P(k;1.67)$ is shown in Fig. A.6.

Appendix B

Differential Equations

In working in biology, which is largely a descriptive science, one soon forgets about derivatives, integrals, and differential equations. This is a pity, because even a rudimentary knowledge of applied mathematics can be useful.

Ordinary differential equations

Take, for example, the problem of preparing the sucrose gradient used in the experiment for visualizing the Gaussian distribution described in Chapter 1. There are several ways of preparing such a gradient; one is shown in Fig. B.1. A mixing chamber of volume x_0 (ml) is filled with water, which is displaced gradually by a solution of sucrose at concentration y_0 (mM). The effluent is piped to the bottom of a graduated cylinder. Initially, the mixture is relatively dilute; the solution that emerges early floats on the solution that emerges later. What is the concentration of sucrose, y, as a function of the volume of the effluent, x? To find out, we write down and solve a simple differential equation: we note at an arbitrary point x, y how y changes with x and then find an expression $y = f(x)$ that accounts for this behavior. When an increment of solution of volume dx ml moves through the apparatus, $y_0 dx$ μmoles of sucrose flow into the mixing chamber and $y dx$ μmoles flow out; therefore the concentration in the mixing chamber increases by an increment $dy = (y_0 - y) dx/x_0$ mM. This implies that

$$\frac{dy}{dx} = \frac{y_0 - y}{x_0}. \tag{B.1}$$

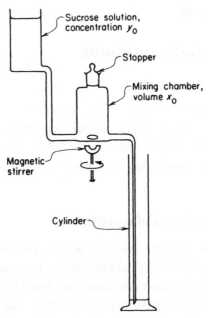

Fig. B.1. Apparatus for preparing a sucrose density gradient. A solution of sucrose flows into a cylinder via a mixing chamber filled initially with water.

This expression tells us how the rate of change of y with respect to x, dy/dx, varies with the concentration, y. This rate is large at the beginning of the experiment when $y = 0$, but it gradually declines as y approaches y_0. The initial slope of the function $y = f(x)$ is y_0/x_0, and the final slope approaches 0. Therefore, we can sketch the solution without having to do the mathematics, as shown in Fig. B.2.

To get the formal solution, we write Eq. B.1 in the form

$$\frac{dx}{x_0} + \frac{dy}{y - y_0} = 0. \tag{B.2}$$

Here, the variables are separated: the term containing dx is a function only of x, and the term containing dy is a

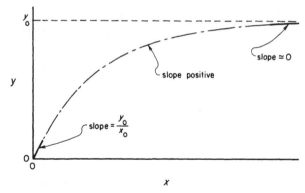

Fig. B.2. The concentration, y, as a function of the volume, x, inferred from inspection of Eq. B.1, given that $y = 0$ at $x = 0$ and $y \to y_0$ as $x \to \infty$.

function only of y. In this case, the solution can be found by integrating each term and equating the sum of the integrals to an arbitrary constant:

$$\frac{x}{x_0} + \ln(y - y_0) = c. \tag{B.3}$$

The constant is determined by the initial (or boundary) values of the problem, e.g., from the constraint $y = 0$ at $x = 0$. Thus $c = \ln(-y_0)$, and

$$y = y_0(1 - e^{-x/x_0}). \tag{B.4}$$

This is the function sketched in Fig. B.2. The gradient is exponential.

The gradient can be prepared in another way if the sucrose reservoir is lowered to the level of the mixing chamber, the stopper is removed, and the level of the liquid in either vessel is allowed to fall at the same rate. A gravity feed will work if the connection between the two vessels is shortened and the flow is restricted at the output. But now the expression for the rate of change of the concentration of sucrose in the mixing chamber, y, with respect to the volume of the effluent, x, is more complicated, because the volume of the solution in the mixing chamber decreases with time. If the cross-sectional areas of the two vessels are the same, half of the solution reaching the cylinder comes from each. Thus, when an increment of solution of volume dx leaves the apparatus $\frac{1}{2}y_0 dx$ μmoles of sucrose flow into the mixing chamber and $y dx$ μmoles flow out. Meanwhile, the volume of the solution in the mixing chamber decreases from $x_0 - \frac{1}{2}x$ to $x_0 - \frac{1}{2}x - \frac{1}{2}dx$. As a result, the concentration of sucrose in the mixing chamber increases by an increment $dy = (\frac{1}{2}y_0 - y)dx/(x_0 - \frac{1}{2}x) + \frac{1}{2}y dx/(x_0 - \frac{1}{2}x)$. The first term is due to the influx of sucrose and is equal to the number of moles added divided by the volume, while the second term is due to the change in volume and is equal to the concentration times the fractional change in volume. (To show this formally, write $y = N/V$, where N is the number of moles of sucrose in the mixing chamber and V is the volume of the solution in the mixing chamber, keeping in mind that both are functions of x, and take the derivative of y with respect to x.) Eq. B.1 now takes the form

$$\frac{dy}{dx} = \frac{y_0 - y}{2x_0 - x}. \tag{B.5}$$

At the beginning of the experiment, $y = 0$, $x = 0$, and the concentration changes at the rate $dy/dx = y_0/2x_0$, half as rapidly as before. As y approaches $\frac{1}{2}y_0$, x approaches $2x_0$, and the rate remains the same. Evidently, the gradient is linear. By separating variables and integrating, as before, we find $y = (y_0/2x_0)x$.

In deriving these expressions, we defined an infinitesimal increment of concentration, dy, and an infinitesimal increment of volume, dx, and wrote the derivative of one with respect to the other as dy/dx. In deriving most of the differential equations that appear in the text, e.g., the equation for the mean time to capture in one dimension, Eq. 3.10, we proceed more systematically, first writing a difference equation for the change in the value of the dependent variable, $y(x + \delta x) - y(x)$, resulting from a finite change in the value of the independent variable, δx, and then passing to the limit

$$\frac{dy}{dx} = \lim_{\delta x \to 0} \frac{y(x + \delta x) - y(x)}{\delta x}. \tag{B.6}$$

Equations B.1 and B.5 are ordinary differential equations; there is one dependent variable (the concentration, y) and one independent variable (the volume, x). They also are first order differential equations; the highest derivative that appears in either equation is the first derivative.

Partial differential equations

Equations 2.3 and 2.5 are partial differential equations; there is more than one independent variable. Equation 2.3 has two independent variables (the position, x, and the time, t), and Eq. 2.5 has four (the position, expressed as x, y, z in cartesian coordinates or as r, θ, ϕ in spherical coordinates, and the time, t). If y is a function of x and t, then

$$\frac{\partial y}{\partial x} = \lim_{\delta x \to 0} \frac{y(x + \delta x, t) - y(x, t)}{\delta x}, \tag{B.7}$$

and

$$\frac{\partial y}{\partial t} = \lim_{\delta t \to 0} \frac{y(x, t + \delta t) - y(x,t)}{\delta t}. \quad \text{(B.8)}$$

In the first of these expressions, t is constant; in the second, x is constant. In a partial derivative, all the independent variables are held constant except the one that is specified in the derivative. Equations 2.3 and 2.5 are second-order differential equations; the highest derivative that appears is the second derivative. All of these equations are linear (of first degree); the highest power in which the dependent variable and its derivatives appear is the first power. Also, all of their terms have constant coefficients. A number of examples of their solutions are given in the text.

In dealing with partial differential equations, it is best to simplify the geometry and reduce the number of independent variables. This is relatively easy when dealing with problems in biology, because subtleties of geometry rarely matter. Suppose, for example, that we want to know the maximum rate of uptake of oxygen by a human red blood cell suspended in a large volume of plasma containing oxygen at concentration C_0. If we ignore the fact that the cell is a biconcave disk and assume, instead, that it is a sphere of about the same size, then the concentration of oxygen in its vicinity will depend on the radius, r, but not on other positional coordinates, θ and ϕ. In this case, an equation with four independent variables (Eq. 2.5) reduces to one with two independent variables (Eq. 2.7); in the steady state, this reduces, in turn, to an ordinary differential equation (Eq. 2.17), which has a trivial solution (Eq. 2.18). As discussed in the text, the maximum rate of uptake attainable by diffusion is sensitive to the size but not to the shape of the adsorber. The solution for the biconcave disk differs from that for the sphere by less than a factor of 2. One rarely knows enough about other variables that determine the behavior of a biological system to justify working to an accuracy much better than this. Extensive discussions of the differential equations that appear in the text can be found in books on classical mathematical physics. Look up the heat or diffusion equation (Eq. 2.5), Laplace's equation (Eq. 2.16), and Poisson's equation (Eq. 3.14). See also books that deal exclusively with the heat equation (Carslaw and Jaeger, 1959) or the diffusion equation (Crank, 1975; Jost, 1960).

Numerical solutions

In the event that one needs to solve a differential equation that is intractable, a solution can be found numerically with a computer. One way of doing this is to go back to the difference equation from which the differential equation was derived and work out the values of the dependent variable on a set of lattice points, step by step. For problems involving diffusion, this is tantamount to a simulation of the random walk. If only the steady-state solution is required, one can guess initial values that approximate this solution and let the system relax to its final state.

As an example, suppose that we prepare a gradient by layering 7 volumes of water over 3 volumes of a solution of sucrose at concentration y_0. What does the gradient look like? To find out, we divide the volume coordinate, x, into a number of lattice points, say 40, and assign an initial concentration 0 to the first 28 and y_0 to the remain-

132 — Appendix B

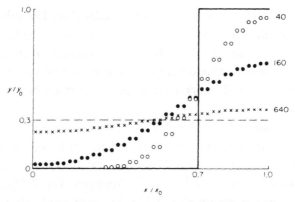

Fig. B.3. The concentration, y, as a function of the volume, x, for a gradient formed by layering $0.7x_0$ ml of water over $0.3x_0$ ml of sucrose at concentration y_0 (solid line). In time, the concentration relaxes to $0.3y_0$ (dashed line). Intermediate values were computed on a lattice of 40 points with 40, 160, and 640 iterations, each corresponding to a time interval $\delta^2/2D$, where δ is the distance between lattice points and D is the diffusion coefficient of sucrose.

ing 12, as shown by the solid line in Fig. B.3. We let each cycle of a do-loop in a computer program correspond to a time interval $\tau = \delta^2/2D$ sec, where δ is the distance in cm between adjacent lattice points and D is the diffusion coefficient of sucrose. This ensures that the system evolves with time as required by Eq. 1.10. The sucrose molecules execute a random walk. During each cycle, half of the molecules at each lattice point step up (to the left in Fig. B.3) and half step down (to the right in Fig. B.3). The boundaries at the top (and bottom) of the gradient are reflecting; molecules at these lattice points that otherwise would step up (or down) remain where they are. This is equivalent to assigning to lattice point i at step n the mean value of the concentrations at the adjacent lattice points $(i - 1)$ and $(i + 1)$ found after step $(n - 1)$, with lattice point 1 (or 40) assigned the mean value of the concentrations at lattice points 1 and 2 (or 39 and 40). The results of such a computation are shown in Fig. B.3 for $n = 40, 160$, and 640. It is evident from the figure that a reasonably steep gradient is obtained that spans the full length of the column of liquid, 40δ, but only for an interval of time extending from about 100τ to 500τ. Thus, a gradient can be set up in this way, provided that one is willing to wait. If $40\delta = 4$ cm and $D = 5 \times 10^{-6}$ cm^2/sec, 100τ is about 1 day.

Computations of this kind can be made in cartesian coordinates in three dimensions by assigning to lattice point (i,j,k) at step n the mean value of the dependent variable at the six adjacent lattice points $(i \pm 1, j \pm 1, k \pm 1)$ found after step $(n - 1)$, with assignments at each boundary determined by whether the boundary is a source, an adsorber, or a reflecting barrier. In two dimensions, one deals similarly with the lattice point (i,j) and the four adjacent lattice points $(i \pm 1, j \pm 1)$. To learn more about numerical solutions, see Smith (1978).

Appendix C
Addendum to Chapter 6

Strategies for chemotaxis

E. coli has receptors for oxygen and other electron acceptors, sugars, amino acids, and dipeptides. It monitors the occupancy of these receptors as a function of time. The probability that a cell will run (rotate its flagella counterclockwise) rather than tumble (rotate its flagella clockwise) depends on the time rate of change of receptor occupancy (Block et al., 1983). We know from responses of cells to short pulses of chemicals delivered by micropipettes (Eq. 2.9) that this measurement spans some 4 sec. A cell compares the occupancy of a given receptor measured over the past second—the aspartate receptor is the only receptor that has been studied in detail—with that measured over the previous 3 sec and responds to the difference (Segall et al., 1986). Now, given rotational diffusion, *E. coli* wanders off course about 60° in 4 sec; see pp. 83–84. If measurements of differences in concentration took much longer than this, they would not be relevant, because cells would change course before the results could be applied. On the other hand, if these measurements were made on a much shorter time scale, their precision would not be adequate. *E. coli* counts molecules as they diffuse to its receptors, and this takes time (Berg and Purcell, 1977). The relative error (the standard deviation divided by the mean) decreases with the square root of the count (pp. 90–91). Thus, in deciding whether life is getting better or worse, *E. coli* uses as much time as it can, given the limit set by rotational Brownian movement.

It is instructive to ask whether *E. coli* might use a simpler strategy for finding regions that it deems favorable. For example, why does it not just tumble more frequently in regions in which the concentration of attractant is high? Would this not serve as a trap? This is a common misconception. Unless the cells tumble incessantly, so that their swimming speed is significantly reduced, the answer is no. If all that the cells do is modulate their turning frequencies on the basis of local cues, their equilibrium distribution is uniform. To see this, go back to the derivation of Fick's first equation, pp. 17–18, and assume that the step lengths, δ, and the step intervals, τ, vary slowly as a function of x. Then, instead of the third equation on p. 18 one obtains

$$J_x \simeq -\frac{1}{2}\left[\frac{\delta_2}{\tau_2} C(x+\delta) - \frac{\delta_1}{\tau_1} C(x)\right],$$

where δ_1 and τ_1 are the step length and step interval that characterize the region near x, and δ_2 and τ_2 are the step length and step interval that characterize the region near $x + \delta$. If the velocity of the cell is constant, so that $\delta_1/\tau_1 = \delta_2/\tau_2$, these terms factor out, and the derivation proceeds as before, yielding $J_x = -D(x) \, \partial C/\partial x$, with D a function of x. The equilibrium distribution remains uniform: at equilibrium, $J_x = 0$, $\partial C/\partial x = 0$, and C is constant, whether D is constant or not. However, spatial variation in D will affect the approach to equilibrium. For example, suppose we release cells at the center of a box containing a source for a chemical at one wall and a sink for that chemical at the opposite wall, using a chemical that causes the cells to tumble more frequently. Then the cells will reach the sink before they reach the source; however, given more time, their distribution will be uniform.

The situation is more complicated if $\delta_1/\tau_1 \neq \delta_2/\tau_2$; for example, if δ varies and τ is constant or if δ is constant and τ varies: a different flux equation results in either case. If δ, τ and δ/τ all vary, a flux equation cannot be formulated in terms of a diffusion coefficient. However, in all cases, the equilibrium cell density varies inversely with the speed, δ/τ; see Schnitzer et al. (1990).

Movement of cells across artificial membranes

Suppose we measure the rate at which bacteria pass through a porous membrane separating two stirred chambers, in an apparatus of the kind shown in Fig. C.1. We add a suspension of bacteria to chamber 1 and measure the flux of cells (the number per cm^2 per sec) that cross the membrane and enter chamber 2. A suitable membrane for this purpose is a glass plate comprising a fused array of capillary tubes (a microchannel plate of the sort used in image intensifiers). A convenient way to

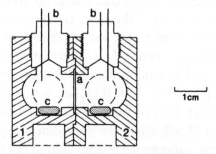

Fig. C.1. Two stirred chambers, 1 and 2, separated by a porous membrane, a. Fluid can be added or removed through pairs of tubes in caps, b. To minimize pressure variations that might force fluid through the plate, the magnetic stir bars, c, counterrotate synchronously. The recess beneath each stir bar is for its driving magnet (not shown). The dashed circles indicate windows in front and behind through which one can view the contents of either chamber. From Berg and Turner (1990).

monitor cell densities is by light scattering. In practice, both the diameter of the tubes and the cell densities can be so small that the chance of two bacteria being in the same tube at the same time is relatively small, so the cells move independently of one another. Since there is a source and a sink, a linear gradient of cells will be set up across the membrane, as shown in Fig. 2.2. This gradient will be established in a time span of order $t = d^2/2D$, where d is the thickness of the membrane (the length of the tubes), and D is the diffusion coefficient of the bacteria (Eq. 1.10). Since the cells interact with the walls of the capillary tubes, this coefficient is bound to differ somewhat from that for cells moving in free space; however, we ignore this complication. For $d = 0.05$ cm and $D \simeq 4 \times 10^{-6}$ cm^2/sec (wild type *E. coli,* p. 93), $t \simeq 5$ min. For the next hour or so, until a significant fraction of the cells have crossed the membrane, the flux will be given by $J_0 = C_0 D/d$, where C_0 is the number of bacteria per unit volume initially present in chamber 1 (Fig. 2.2). Thus, D can be determined experimentally from measurements of J_0.

Now suppose we add a chemical attractant to chamber 2. Over a time span of about 5 min, a linear gradient of the attractant will be set up in the tubes, and the bacteria will drift toward chamber 2. As a result, the flux of bacteria crossing the membrane will increase from J_0 to some larger value J_+. If, instead, we add the attractant to chamber 1, the flux will decrease to some smaller value, J_-. Steady-state values for J_+ and J_- can be computed from Eq. 4.4 if one assumes that the bacterial drift velocity, $v = v_d$, is constant. Integrate this equation to find C as a function of x, holding J and v constant. Then apply the boundary conditions $C = C_0$ at $x = 0$ and $C = 0$ at $x = d$, and solve for J; see p. 923 of Berg and Turner (1990).

The results are

$$J_+ = C_0 v e^\alpha / (e^\alpha - 1), \quad (C.1)$$

and

$$J_- = C_0 v / (e^\alpha - 1), \quad (C.2)$$

where v is the magnitude of the drift velocity (always positive), and

$$\alpha = vd/D. \quad (C.3)$$

Note that $J_+/J_- = e^\alpha$. Thus, v can be determined experimentally from measurements of J_+, J_- or J_+/J_-.

If α is small, $e^\alpha \simeq 1 + \alpha$, and both J_+ and J_- approach $C_0 v/\alpha = J_0$, as expected. Diffusion with drift (chemotaxis) can be distinguished from diffusion alone only if α is large. For *E. coli*, this is the case only if the membrane is thick, not thin. For $d = 0.05$ cm, $v \simeq 3 \times 10^{-4}$ cm/sec, and $D \simeq 4 \times 10^{-6}$ cm^2/sec (p. 93), $\alpha \simeq 4$, and $J_+/J_- \simeq 55$. A simpler way to see that the membrane must be thick is to note that the time required for a cell to cross the membrane by drift is d/v, while the time required for a cell to cross the membrane by diffusion is of order $d^2/2D$. If one wants to study chemotaxis, transit by diffusion should take longer than transit by drift. Thus, $d^2/2D > d/v$, or $d > 2D/v$.

The assumption that the drift velocity, v, is constant is only approximate, because cells respond to the time rate of change of receptor occupancy (p. 134). For a cell swimming at a constant velocity in a linear gradient, the drift velocity decreases once the cells enter regions in which the concentration of the attractant is larger than the dissociation constant of the receptor. The cells become less sensitive to the attractant as the receptors saturate.

Appendix C—139

Movement of ions across cell membranes

Neurophysiologists have seen Eqs. C.1–C.3 in a different context. Instead of bacteria, consider ions, such as Na^+, K^+, or Cl^-, drifting across a cell membrane in response to a constant electric field, V/d. Here, V is the electrical potential difference across the membrane (chamber 1 less chamber 2). Recall from our discussion of electrophoresis, pp. 62–64, that the force exerted by this field on an ion of charge q is approximately $(V/d)q$. This force is balanced by viscous drag: from Eqs. 4.2 and 4.3, the drift velocity for the ion is $v = (V/d)q/(kT/D) = (Vq/kT)(D/d)$, where D is the diffusion coefficient for the ion. The magnitude of the drift velocity is directly proportional to the diffusion coefficient, and α, Eq. C.3, is just Vq/kT. It is the same, except for sign, for all univalent ions.

Suppose we add different concentrations of NaCl and KCl to chambers 1 and 2 and use a membrane that is permeable to any ion. The ions will cross the membrane at different rates, depending on their concentration gradients and diffusion coefficients. Net displacement of charge across the membrane will generate an electrical potential difference, and this will cause the ions to drift. The electric field will retard the displacement of the dominant ion, and the system will come to a steady state. What is the steady-state transmembrane potential? To find out, write down the net flux for each ion, using Eqs. C.1 and C.2, keeping in mind that the ions respond to the same electric field and that positive and negative ions drift in opposite directions. If the charge distributions in the two chambers are to remain constant, the net flux of positive ions moving from chamber 1 to chamber 2 (the number moving from 1 to 2 less the number moving from

2 to 1) must be equal to the net flux of negative ions moving in the same direction. So, write down this equality and solve for e^α. The problem is simpler than it looks (for univalent ions), because ($e^\alpha - 1$) factors out. Denote the concentration of Na^+ in chamber 1 by $[Na]_1$ and the magnitude of the drift velocity for Na^+ by v_{Na}, etc. Then

$$e^\alpha = \frac{v_{Na}[Na]_2 + v_K[K]_2 + v_{Cl}[Cl]_1}{v_{Na}[Na]_1 + v_K[K]_1 + v_{Cl}[Cl]_2}.$$

Finally, take the logarithm of either side and rearrange:

$$V = \frac{kT}{q} \ln \frac{v_{Na}[Na]_2 + v_K[K]_2 + v_{Cl}[Cl]_1}{v_{Na}[Na]_1 + v_K[K]_1 + v_{Cl}[Cl]_2}. \quad (C.4)$$

Since the drift velocities are proportional to the diffusion coefficients with one and the same proportionality constant, this equation could be expressed equally well with Ds instead of vs. A more general solution is obtained if we recognize that different ions might use different channels of distinct size and number. In this case, apply different scale factors to the flux equations for each ion, Eqs. C.1 and C.2, and note that these factors will appear in front of the corresponding vs in Eq. C.4. These products are called permeability coefficients, P_{Na}, P_K, and P_{Cl}. When written in this form, Eq. C.4 is known as the Goldman equation (Goldman, 1943; Hodgkin and Katz, 1949).

Suppose, as in a resting axon, that the concentration of Na^+ inside the cell (chamber 1) is smaller than its concentration outside (chamber 2), that the concentration of K^+ is larger inside the cell than outside, and that the membrane is more permeable to K^+ than to either Na^+ or Cl^-. Then the transmembrane potential (inside less outside) is negative, as expected. If the membrane is permeable only to K$^+$, Eq. C.4 reduces to the Nernst potential:

$$V = \frac{kT}{q} \ln \frac{[K]_2}{[K]_1} \qquad (C.5)$$

This is an equilibrium rather than a steady-state potential. It can be written down at once from the Boltzmann equation, Eq. 5.3, given that the energy gained when an ion crosses the membrane from side 1 to 2 is $-qV$. In this limit, potassium ions continue to move back and forth across the membrane, but their net flux is zero.

A reader who knows more physics will realize that our derivation of Eq. C.4 is not self-consistent. If the electric field in the channel is constant, as we have assumed, there will be substantial differences in the concentrations of positive and negative ions at different points along the channel, just as there would be for two different kinds of bacteria responding to a chemical that attracts one and repels the other; see Fig. 3 of Berg and Turner (1990). This charge imbalance will perturb the field. The derivation is suspect unless any net charge density can be neutralized. Goldman recognized this problem. In order to obtain a simple solution, he assumed that his membranes (mostly mixtures of collodion and cephalin) contained a large number of dipolar ions near their isoelectric point that would shield such charge. The problem is less severe in real cell membranes, where narrow aqueous channels are sparsely distributed in a lipid bilayer that has a large electrical capacitance. Most of the free charge is at the surface of the membrane, and only a small fraction is in channels.

Appendix D

Constants and Formulas

The following constants and formulas are useful for order-of-magnitude calculations. They are worth committing to memory. They are given in cgs units.

kT — The energy of thermal fluctuation, where k is Boltzmann's constant and T is the absolute temperature: 4×10^{-14} g cm^2/sec^2 (ergs) at room temperature.

D_w — The diffusion coefficient of water in water at room temperature: 10^{-5} cm^2/sec.

η_w — The coefficient of viscosity of water at room temperature: 10^{-2} g/cm sec (poise).

$6\pi\eta a v$ — The viscous drag (in g cm/sec^2 = dynes) on a sphere of radius a (cm) moving at velocity v (cm/sec) through a medium of viscosity η.

$D = kT/6\pi\eta a$ — The translational diffusion coefficient of a sphere of radius a in a medium of viscosity η.

$D_r = kT/8\pi\eta a^3$ — The rotational diffusion coefficient of a sphere of radius a in a medium of viscosity η.

$\langle x^2 \rangle = 2Dt$ — The mean-square displacement (cm^2) in each dimension of a particle of translational diffusion coefficient D in time t (sec).

$\langle \theta^2 \rangle = 2D_r t$ The mean-square angular displacement (radians^2) about an axis of a particle of rotational diffusion coefficient D_r in time t.

$N(x) = N(0)e^{-[E(x)-E(0)]/kT}$ The equilibrium distribution of numbers of particles $N(x)$ in states of energy $E(x)$.

Bibliography

Abramowitz, M., and Stegun, I. A., eds. 1972. *Handbook of Mathematical Functions*. National Bureau of Standards. Also reprinted by Dover.

Berg, H. C. 1978. The tracking microscope. *Advances in Optical and Electron Microscopy* **7**, 1–13.

Berg, H. C., and Brown, D. A. 1972. Chemotaxis in *Escherichia coli* analysed by three-dimensional tracking. *Nature* **239**, 500–504.

Berg, H. C., and Purcell, E. M. 1967a. A method for separating according to mass a mixture of macromolecules or small particles suspended in a fluid, I. Theory. *Proceedings of the National Academy of Sciences USA* **58**, 862–869.

Berg, H. C., and Purcell, E. M. 1967b. A method for separating according to mass a mixture of macromolecules or small particles suspended in a fluid, III. Experiments in a centrifugal field. *Proceedings of the National Academy of Sciences USA* **58**, 1821–1828.

Berg, H. C., and Purcell, E. M. 1977. Physics of chemoreception. *Biophysical Journal* **20**, 193–219.

Berg, H. C., Stewart, W. W., and Purcell, E. M. 1967. A method for separating according to mass a mixture of macromolecules or small particles suspended in a fluid, II. Experiments in a gravitational field. *Proceedings of the National Academy of Sciences USA* **58**, 1286–1291.

Berg, H. C., and Turner, L. 1979. Movement of microorganisms in viscous environments. *Nature* **278**, 349–351.

Berg, H. C., and Turner, L. 1990. Chemotaxis of bacteria in glass capillary arrays. *Biophysical Journal* **58**, 919–930.

Berg, H. C., and Turner, L. 1991. Selection of motile nonchemotactic mutants of *Escherichia coli* by field flow fractionation. *Proceedings of the National Academy of Sciences USA* **88**, 8145-8148.

Block, S. M., Segall, J. E., and Berg, H. C. 1983. Adaptation kinetics in bacterial chemotaxis. *Journal of Bacteriology* **154**, 312-323.

Carslaw, H. S., and Jaeger, J. C. 1959. *Conduction of Heat in Solids*. Oxford, 2d ed.

Crank, J. 1975. *The Mathematics of Diffusion*. Oxford, 2d ed.

Davenport, W. B., and Root, W. L. 1958. *An Introduction to the Theory of Random Signals and Noise*. McGraw-Hill.

DeLisi, C., and Wiegel, F. W. 1981. Effect of non-specific forces and finite receptor number on rate constants of ligand—cell bound-receptor interactions. *Proceedings of the National Academy of Sciences USA* **78**, 5569-5572.

Dwight, H. B. 1961. *Tables of Integrals and Other Mathematical Data*. Macmillan, 4th ed.

Feynman, R. P., Leighton, R. B., and Sands, M. 1963. *The Feynman Lectures on Physics*, Vol. 1. Addison-Wesley.

Futrelle, R. P. 1984. How molecules get to their detectors: the physics of diffusion of insect pheromones. *Trends in NeuroSciences* **7**, 116-120.

García de la Torre, J., and Bloomfield, V. A. 1981. Hydrodynamic properties of complex, rigid, biological macromolecules: theory and applications. *Quarterly Reviews of Biophysics* **14**, 81-139.

Giddings, J. C. 1988. Field-flow fractionation. *Chemical and Engineering News* **66**(41), 34-45.

Goldman, D. E. 1943. Potential, impedance, and rectification in membranes. *Journal of General Physiology* **27**, 37–60.

Hodgkin, A. L., and Katz, B. 1949. The effect of sodium ions on the electrical activity of the giant axon of the squid. *Journal of Physiology* **108**, 37–77.

Jaeger, J. C. 1965. Diffusion from constrictions. In *Studies in Physiology*, D. R. Curtis and A. K. McIntyre, eds., pp. 106–117. Springer-Verlag.

Jost, W. 1960. *Diffusion in Solids, Liquids, Gases*. Academic.

Kirkland, J. J., and Yau, W. W. 1982. Sedimentation field flow fractionation: applications. *Science* **218**, 121–127.

Lamb, H. 1932. *Hydrodynamics*. Cambridge, 6th ed. Also reprinted by Dover.

Landau, L. D., and Lifshitz, E. M. 1959. *Fluid Mechanics*. Addison-Wesley.

Lovely, P. S., and Dahlquist, F. W. 1975. Statistical measures of bacterial motility and chemotaxis. *Journal of Theoretical Biology* **50**, 477–496.

Macnab, R. M., and Ornston, M. K. 1977. Normal-to-curly flagellar transitions and their role in bacterial tumbling. Stabilization of an alternative quartenary structure by mechanical force. *Journal of Molecular Biology* **112**, 1–30.

Meidner, H., and Mansfield, T. A. 1968. *Physiology of Stomata*. McGraw-Hill.

Perrin, F. 1934. Mouvement Brownien d'un ellipsoïde (I). Dispersion diélectrique pour des molécules ellipsoïdales. *Le Journal de Physique et le Radium*, Série 7, **5**, 497–511.

Perrin, F. 1936. Mouvement Brownien d'un ellipsoïde (II). Rotation libre et dépolarisation des fluorescences. Translation et diffusion de molécules ellipsoïdales. *Le Journal de Physique et le Radium*, Série 7, **7**, 1-11.

Purcell, E. M. 1977. Life at low Reynolds number. *American Journal of Physics* **45**, 3-11.

Schnitzer, M. J., Block, S. M., Berg, H. C., and Purcell, E. M. 1990. Strategies for chemotaxis. In *Biology of the Chemotactic Response*, ed. J. P. Armitage and J. M. Lackie, pp. 15-34. Society for General Microbiology Symposium, Vol. 46. Cambridge University Press.

Segall, J. E., Block, S. M., and Berg, H. C. 1986. Temporal comparisons in bacterial chemotaxis. *Proceedings of the National Academy of Sciences USA* **83**, 8987-8991.

Shoup, D., and Szabo, A. 1982. Role of diffusion in ligand binding to macromolecules and cell-bound receptors. *Biophysics Journal* **40**, 33-39.

Smith, G. D. 1978. *Numerical Solution of Partial Differential Equations*. Oxford, 2d ed.

Smythe, W. R. 1950. *Static and Dynamic Electricity*. McGraw-Hill, 2d ed.

Zwanzig, R. 1990. Diffusion-controlled ligand binding to spheres partially covered by receptors: an effective medium treatment. *Proceedings of the National Academy of Sciences USA* **87**, 5856-5857.

Index

adsorber, diffusion to: disk-like, 27–28, 30–33; ellipsoidal, 29; planar, 18–19, 41–46; spherical, 26–27, 38–41, 47
agarose, 64
ampholytes, 74
aperture, circular, diffusion through, 28, 34–36
autocorrelation, 105–110
axon, 139–141

bacterium: chemotaxis of, 94, 134–138; diffusion of, 93–94, 134–138; inability to coast, 76–77; propulsion of, 78–81; Reynolds number of, 75–76
barrier, reflecting, planar, 34–36, 43–45
binomial: coefficient, 118; distribution, 13, 118–121
Boltzmann: constant, 5, 142; distribution, equation, 66–70, 72, 102–104, 143
Bose-Einstein statistics, 70
Brownian movement, 5, 77, 134
bubble, viscous drag on, 56
buoyancy, 58–59, 73–74

capacitance, electrical, 29
capture: mean time to, 42–46; probability of, 38–42, 46–47
cells: diffusion through pores between, 35; diffusion to receptors on, 30–33; movement across membranes, 136–138; sedimentation of, 56, 62
centrifuge, 58–62, 72–74
cesium chloride, 64, 73
chemotaxis: bacterial, 94, 134–138; strategies for, 134–136
chromatography: autocorrelation analysis of, 106–109; gas,

high-pressure liquid, partition, 100–101
coefficient, *see* diffusion coefficient; frictional drag coefficient; partition coefficient; viscosity coefficient
compound events, 112–113
convective flow, 15–16, 60, 63–64
correlation time, 50, 109
countercurrent distribution, 95–100

density gradient, 15, 60–61, 73–74, 125–128, 131–133
differential equations: numerical solutions to, 131–133; ordinary, 125–129; partial, 129–131. *See also* diffusion equation; heat equation; Laplace equation; Poisson equation
diffusion: bacterial, 93–94; current, 27–36, 39, 41–42; in a Boltzmann distribution, 67–68, 103–104; in a pipe, 23–25, 130–133; from a point source, 22–23; through a circular aperture, 28, 34–36; to an adsorbing disk, 27–28, 30–33, 36; to an adsorbing ellipsoid, 29; to an adsorbing sphere, 26–27, 38–41, 47; with drift, 48–51, 94, 134–138
diffusion coefficient: of disk, ellipsoid, 56–58; of small molecule in air, 16; of small molecule in water, 10, 16, 142; of sphere, 56, 83, 142; rotational, definition of, 81–83; translational, definition of, 10, 18–19
diffusion equation: derivations,

diffusion equation (cont.)
17–21, 50–51; including drift,
50–51; steady-state solutions
of, 25–36, 38–39, 41–42;
time-dependent solutions of,
21–25, 131–133
diffusion resistance, 31–36
disk: diffusion to, 27–28, 30–33,
36; viscous drag on, 56–58,
84–85
distribution, see binomial distribution; Boltzmann distribution;
Gaussian distribution; Poisson
distribution; Poisson interval
distribution
DNA: diffusion of repressor along,
44–45; separation of, 64, 74
drag, see viscous drag
drift, random walk with, 48–50,
94, 137–138
droplet, viscous drag on, 56

effective mass: definition of, 58;
of a sphere, 62
Einstein-Smoluchowski relation,
49, 71–72
electrical analogue: for diffusion
current, 29, 31–34; for mean
time to capture, 46
electrical potential, across membrane, 139–141
electrophoresis, 62–64, 74
ellipsoid: diffusion to, 29; viscous
drag on, 56–58, 84–85
equilibrium: distribution, 65–70,
135–136; sedimentation, 72–74
error function, 22–23
escape, see capture
Escherichia coli, see chemotaxis;
motility
expectation value, *see* mean value

Fermi-Dirac statistics, 70
Fick's equation, *see* diffusion
equation

fish, Reynolds number of, 76
flagellum, *see* propulsion
flow, *see* viscous flow
flux: of bacteria, 135–138; of diffusing particles, 17–21; of
ions, 139–141
frictional drag coefficient, rotational: definition of, 82; of
disk, ellipsoid, sphere, 83–85
frictional drag coefficient, translational: definition of, 49; of bubble, disk, droplet, ellipsoid,
sphere, 56–58

Gaussian distribution, 14–16, 22,
73, 121–123
gel, 64
Goldman equation, 140

heat equation, 21

ion: electrophoresis of, 62–64,
74; isoelectric focusing of, 74;
movement across membrane of,
139–141
isoelectric focusing, 74

kinetic energy, 5, 71, 81
kT, 5, 70–71, 81, 142

Laplace equation, 35, 47, 131
Laplacian, 21
leaves, *see* stomata
lysozyme, 5, 14, 58, 101

mean value: definition of,
114–115; of velocity of molecules during chromatography,
99, 105, 107. *See also* binomial
distribution; Gaussian distribution; Poisson distribution;
Poisson interval distribution
mean-square: angular displacement, 82, 143; displacement,
9–10, 142; velocity, 5, 71. *See*

also root-mean-square
membrane: movement of bacteria
across, 136–138; movement of
ions across, 139–141; potential, 139–141
methylcellulose, 53
micropipette, as a point source, 22–23
motility, bacterial, 75–86, 91–94, 134–138
mutually exclusive outcomes, 111

Nernst potential, 141
normal curve of error, 122–123
normal distribution, *see* Gaussian distribution
numerical methods for solving differential equations, 131–133. *See also* simulation

ordinary differential equations, *see* differential equations

partial differential equations, *see* differential equations
partition: chromatography, 100–101; coefficient, 96
pH gradient, 74
pheromones, detection of, 34
pipe: diffusion in, 23–25, 131–133; flow through thin rectangular, 53–54
plating bacteria, 90
point source, diffusion from, 22–23
poise, units, 52
Poisson: distribution, 88–89; 123–124; equation, 46, 131; interval distribution, 87–88
polyacrylamide, 64, 100
power, to push cell, 77
pressure, driving flow, 53–54
probability: classical or a priori, 111; conditional, 112; distributions, 113–118; of capture, 37–42, 46–47; machine, 117
propulsion, flagellar, 78–79

radioactive decay, 90–91
random changes in direction, 85–86, 93–94
random walk: biased, 50, 94, 97, 134–138; computer simulation of, 11–12, 68, 91–92, 104; in one dimension, 6–7, 48–50, 81–82; in two, three dimensions, 11–12, 81, 83–84, 91–92; rotational, 81–82; step length, 14; step rate, 14; with drift, 48–50
receptors: bacterial, 134; diffusion to, 30–33
resistance, diffusion, 31–36
Reynolds number, 75–76
ribosome, 59, 63
root-mean-square: displacement, 9–10; velocity, 5, 60. *See also* mean-square
rotational diffusion, 81–85
runs, bacterial, 80–81, 86, 91–94, 134–136

S, units, 59
scale height, 67, 103
Schlieren optics, 61
sedimentation: equilibrium, 65–68, 72–74, 102–103; field-flow fractionation, 102–105; rate, 58–62
separation: by countercurrent distribution, 95–100; by density-gradient sedimentation, 73–74; by electrophoresis, 62–64; by isoelectric focusing, 74; by partition chromatography, 100–101, 106–109; by sedimentation field-flow fractionation, 102–105; by sedimentation rate, 58–62; criterion for, 97–

separation (*cont.*)
  98; of variables, 126–127
simulation: of bacterial motion, 91–92; of diffusion in Boltzmann distribution, 68; of diffusion in field-flow fractionation, 104; of diffusion in finite pipe, 131–133; of elution profiles in field-flow fractionation, 105; of Poisson process, 91; of random walk, 11–12, 68, 91–92, 104; of successive trials, 117
sodium dodecyl sulfate, 64
source, diffusion from: planar, 18–19, 34–35, 41–42; point, 22–23; spherical, 38–39
sphere: adsorbing, 26–27, 38–41, 47; diffusion coefficient of, 56, 83, 142; effective mass of, 62; probability of capture by, 38–41, 47; sedimentation rate of, 62; viscous drag on, 55–56, 83, 142; with disk-like adsorbers, 30–33
standard deviation: definition of, 115; of velocity of molecules during chromatography, 99, 106–108. *See also* binomial distribution; Gaussian distribution; Poisson distribution; Poisson interval distribution

statistical independence, 112
stomata, diffusion through, 36
stoke, units, 52
Stokes' law, 55
sucrose gradient, 15, 60–61, 64, 125–128, 131–133
Svedberg equation, 59

torque: on a flagellar filament, 78–79; on a particle in a viscous medium, 82; on a swimming cell, 78
tumbles, bacterial, 80–81, 86, 91–94, 134–136

variance, *see* standard deviation
viscosity coefficient: definition of, 52; of air, water, glycerol, 52–53
viscous drag: on a bubble, 56; on a disk, 56–58, 84–85; on a droplet, 56; on a flagellar filament, 78–79; on a sphere, 55–56, 83, 142; on a swimming cell, 78; on an ellipsoid, 56–58, 84–85
viscous flow: around a sphere, 54–55; basic equation for, 54; through a thin rectangular channel, 53–54
viscous shear, 51–53, 55, 75

Printed and bound by CPI Group (UK) Ltd, Croydon, CR0 4YY
29/04/2025
14663614-0001